HARDWARE-SOFTWARE
CO-SYNTHESIS OF
DISTRIBUTED EMBEDDED
SYSTEMS

HARDWARE-SOFTWARE CO-SYNTHESIS OF DISTRIBUTED EMBEDDED SYSTEMS

Ti-Yen YEN
Quickturn Design Systems
Mountain View, California, USA

Wayne WOLF
Princeton University
Princeton, New Jersey, USA

KLUWER ACADEMIC PUBLISHERS

BOSTON / DORDRECHT / LONDON

A C.I.P. Catalogue record for this book is available from the Library of Congress

ISBN 978-1-4419-5167-0

Published by Kluwer Academic Publishers,
P.O. Box 17, 3300 AA Dordrecht, The Netherlands.

Kluwer Academic Publishers incorporates
the publishing programmes of
D. Reidel, Martinus Nijhoff, Dr W. Junk and MTP Press.

Sold and distributed in the U.S.A. and Canada
by Kluwer Academic Publishers,
101 Philip Drive, Norwell, MA 02061, U.S.A.

In all other countries, sold and distributed
by Kluwer Academic Publishers Group,
P.O. Box 322, 3300 AH Dordrecht, The Netherlands.

Printed on acid-free paper

CONTENTS

PREFACE   vii

1 INTRODUCTION   1

    1.1 Overview   1

    1.2 Hardware-Software Co-Synthesis Problems   2

    1.3 Characteristics of Embedded Systems   4

    1.4 Outline of Book   9

2 PREVIOUS WORK   13

    2.1 Uniprocessor Scheduling and Performance Analysis   13

    2.2 Scheduling Real-Time Distributed System   18

    2.3 Co-Specification and Co-Simulation   23

    2.4 Hardware-Software Partitioning   27

    2.5 Distributed System Co-Synthesis   33

    2.6 Event Separation Algorithms   36

3 SYSTEM SPECIFICATION   41

    3.1 Task Graph Model   41

    3.2 Embedded System Architecture   48

    3.3 The Bus Model   51

4 PERFORMANCE ANALYSIS   57

    4.1 Overview   57

    4.2 Fixed-Point Iteration   60

    4.3 Phase Adjustment   62

    4.4 Separation Analysis   68

    4.5 Iterative Tightening   75

4.6   Period Shifting   77
4.7   Task Pipelining   80
4.8   Experimental Results   81

5   SENSITIVITY-DRIVEN CO-SYNTHESIS   87
5.1   Overview   87
5.2   Sensitivity Analysis   89
5.3   Priority Prediction   92
5.4   Optimization by Stages   95
5.5   The Co-Synthesis Algorithm   101
5.6   Experimental Results   103

6   COMMUNICATION ANALYSIS AND
    SYNTHESIS   117
6.1   Overview   117
6.2   Communication Delay Estimation   118
6.3   Communication Synthesis   128
6.4   Experimental Results   129

7   CONCLUSIONS   137
7.1   Contributions   137
7.2   Future Directions   138

REFERENCES   143

INDEX   155

PREFACE

Embedded computer systems use both off-the-shelf microprocessors and application-specific integrated circuits (ASICs) to implement specialized system functions. Examples include the electronic systems inside laser printers, cellular phones, microwave ovens, and an automobile anti-lock brake controller. With deep submicron technology, an embedded system can be integrated into a single chip and becomes "system on silicon" or "core-based design" for many applications. While such systems have been widely applied in consumer products for decades, the *design automation* of embedded systems has not been well studied. Hand-crafted techniques were sufficient to design small, low-performance systems in the past. Today manual design effort grows rapidly with system complexity and becomes a bottleneck in productivity, so it is important to apply computer-aided design (CAD) methodology for embedded system design.

Embedded computing is unique because it is a co-design problem—the *hardware engine* and *application software architecture* must be designed simultaneously. By hardware engine, we mean a heterogeneous distributed system composed of several processing elements (PE), which are either CPUs or ASICs. By application software architecture, we mean the *allocation* and *scheduling* of processes and communication. New techniques such as *fixed-point iterations*, *phase adjustment*, and *separation analysis* are proposed to efficiently estimate tight bounds on the delay required for a set of multi-rate processes preemptively scheduled on a real-time reactive distributed system. Based on the delay bounds, a gradient-search co-synthesis algorithm with new techniques such as *sensitivity analysis*, *priority prediction*, and *idle-PE elimination* is developed to select the number and types of PEs in a distributed engine, and determine the allocation and scheduling of processes to PEs. New communication modeling is presented to analyze communication delay under interaction of computation and communication, allocate interprocessor communication links, and schedule communication.

Distributed computers are often the most cost-effective means of meeting the performance requirements of an embedded computing application. Most recent work on co-design has focused on only one-CPU architecture. This book is one of the first few attempts to co-synthesize multi-rate distributed embedded systems.

Hardware-software co-design is an important and popular research area, as evidenced by the large attendance of the past several IEEE/ACM/IFIP Hardware-Software Co-Design Workshops. Co-design is of interest to academics and industrial designers. Distributed embedded systems are used in industries including multimedia, defense signal processing, and satellites. This is the first book to describe techniques for the design of distributed embedded systems, which have arbitrary hardware and software topologies—previous work has concentrated on template architectures. This book will be of interest to: academic researchers for personal libraries and advanced-topics courses in co-design; industrial designers who are building high-performance, real-time embedded systems with multiple processors.

1

INTRODUCTION

1.1 OVERVIEW

There are few industries which, like the electronics industry, can continuously improve qualities of products and at the same time lower their prices every year. The growth rate of the electronics industry is about ten times greater than that of the global economy. Computer-aided design methodology for VLSI systems has become indispensable to utilize the complexity of circuits in improving products, while limiting the design cost of systems. Semiconductor physics is not the only technical challenge in moving the technology forward; in order to cope with the more than exponential increase of the design complexity, VLSI design automation is the key to the further advance of the electronics industry.

The contribution of the research in computer-aided design has been demonstrated by the success of several electronic design automation (EDA) companies as well as the in-house CAD groups in major semiconductor companies. Many successful industrial CAD tools were developed according to the trend in raising the levels of abstraction: first in *physical level*, then in *logic level*, and later on in *register-transfer level*. Recently, a few companies have started to promote their products in *behavioral level*, which has been an active research area for almost ten years. In the meantime, the prospect of *system-level* design automation has received a great deal of attention in both academia and industry.

Hardware-software co-design of embedded computer systems [103] is a major problem in system-level design. Hardware-software co-design is not a new field [101, 47], but the computer-aided design support of embedded systems is still primitive. Wolf [101] considers that embedded computer system design is

in a state very similar to VLSI design in the late 1970s: the next 10 years may
be as revolutionary for embedded system design as the last 10 years have been
for IC design.

1.2   HARDWARE-SOFTWARE CO-SYNTHESIS PROBLEMS

In many VLSI systems, hardware and software must be designed together.
Hardware usually can be described by a structural netlist of primitive components such as logic gates. Software means a sequence of code chosen from a given
instruction set for a particular processor. Like the lower levels of abstraction,
a hardware-software co-design problem includes simulation, synthesis, verification, testing, integration, and so on. A hardware-software co-synthesis algorithm can automatically partition a unified specification into hardware and software implementation to meet performance, cost, or reliability goals. Hardware-
software co-design can be applied to several different design problems [68]. In
particular, the following fields have been addressed for co-synthesis in the literature.

1.2.1   ASIP Synthesis

An application-specific instruction processor (ASIP) is a microprocessor with
special micro-architecture and instruction set for a specific domain of programs.
The performance of ASIPs can be improved by designing an instruction set
that matches the characteristics of hardware and the application. As with
ASICs, the low volume production and short turn-around time of ASIPs requires computer-aided design methodology [1, 11, 38, 79].

The design of ASIPs is a hardware-software co-design problem because the
micro-architecture and the instructions are designed together. Given a set of
benchmark programs and the design constraints, an ASIP co-synthesis system synthesizes the datapath and controlpath in the CPU core, as well as
the instruction set for the CPU core. Sometimes, a corresponding compiler or
assembler is also generated for the instruction set. Allocation and scheduling
techniques in high-level synthesis [65] are essential for the synthesis of datapath
and instruction set in ASIPs.

1.2.2   Execution Accelerator

It is usually an accepted concept that for a computer in operation, the under-lying hardware is fixed, and only the software running on it is changeable. The development of Field Programmable Gate Array (FPGA) technology makes programmable hardware a possibility. Given a board of FPGAs connected to a host computer, a portion of a software can be mapped to FPGAs and a speedup over pure software execution can be achieved. Olukotun et al. [71] applied such an approach to accelerate digital system simulation. Emulation technology [99, 64, 50] for design verification of complex systems uses a similar concept.

The hardware-software partition problem for such accelerators is slightly dif-ferent from that in embedded system design or ASIC design with processor cores. The hardware area is considered to be fixed [42] and is not an objective function to minimize; the type of CPU must be predefined.

1.2.3   Core-Based ASIC Design

Due to the rapid growth on the available area of a single chip, it has been possible to integrate processor cores, memory, and surrounding ASIC circuitry into the same chip. Such an implementation is sometimes called a *system on silicon* or *system-level ASIC*, because what used to be called a system can be put on a single integrated circuit. The combination of hardware and software on a chip greatly enhances the functionality of the circuit without increasing the system cost. It may reduce time to market because it is not necessary to re-design the circuitry for the processor core which can be taken from a library.

The code running on the processor core and the ASIC circuitry must be design together. Since it is possible to integrate more than one processor cores into the same chip, the system can be a distributed system. Once the chip is fabricated, reprogramming the processor core is almost impossible, so careful analysis and verification are important. A new hardware-software co-synthesis approach is desirable for the design of such systems.

1.2.4   Embedded System Design

Embedded systems consist of both off-the-shelf microprocessors and application-specific integrated circuits (ASICs) for special applications.

Microprocessor-based systems have made a significant impact on human life since 1970s. The revenue from microcontrollers is larger than that from ASICs in many major semiconductor companies. The microcontroller market is still growing in recent years. A microcontroller is usually much cheaper than ASICs due to mass production. Microprocessors become more and more powerful every year, so pure ASIC systems are unable to replace the ubiquitous microprocessor-based systems. ASICs are often added into embedded systems for time-critical functions which the software performance on microprocessors cannot satisfy. ASICs also help protect intellectual property—a system containing only standard components is easier to copy or imitate by competitors.

Some recent work formulated embedded system synthesis as a hardware-software partitioning problem—partitioning between a single predefined CPU and an ASIC. Such an approach does not model the real design problem. Wolf [103] first proposed the **engine metaphor** to help understand the roles of hardware and software in implementation. A critical decision in the early design stage of an embedded system is the choices of the number and the types of the CPUs—this is reminiscent of engine selection for vehicle design. The right engine cannot be selected for a vehicle without considering its body and aerodynamics. Similarly, the design of software architecture is closely related to hardware engine design.

Although this book concentrates on embedded system co-synthesis, we can share many basic concepts for embedded system design with other hardware-software co-design problems, especially ASIC design with processor cores.

1.3 CHARACTERISTICS OF EMBEDDED SYSTEMS

It is difficult to come up with a comprehensive definition of embedded computing [103]. We are not going to draw a fine line between what are embedded systems and what are not embedded systems. Instead, we will give the features for the co-synthesis problem we will solve in this work. These features do characterize most embedded systems.

1.3.1 Application Specific Systems

Wolf once gave a simple description as "An embedded computer system is a system which uses computers but is not a general-purpose computer." This implies that an embedded system utilizes computer technology for a specific application, and is not as general-purpose as a computer. Because embedded systems are application-specific, their design conditions are different from normal computer design:

- Job characteristics are known before the hardware engine is designed. This allows deterministic performance analysis which helps fine tune the design decisions.

- Reprogramming by the users is impossible after the system design is finished, though some embedded systems use flash memory for upgrades.

Such design conditions naturally lead to different criterion for the design decisions:

- *Static vs. dynamic*: Dynamic scheduling or allocation often causes more overhead in performance and cost than static approaches. Nevertheless, a dynamic scheme is unavoidable or desirable when the jobs to perform on a general-purpose computer is unpredictable. In embedded system, static schemes are often better choices since the job characteristics are known *a priori*.

- *Heterogeneous vs. homogeneous*: Heterogeneous systems have larger design space. However, to ease the reprogramming tasks by either the human or a parallel compiler, general-purpose parallel computers are usually made of homogeneous architectures: the type of CPUs are all the same, and the regular communication topologies such as single-bus connection, point-to-point communication, or mesh structure are adopted. Embedded systems have tight design constraints, and heterogeneity provides better design flexibility.

1.3.2 Hard Real-Time Systems

Many embedded systems are hard real-time systems. For a hard real-time system, the correctness of the system function depends not only on the logical results of computation, but also on the time at which the results are produced. Shin et al. [89], Ramamritham et al. [81], and Sha et al. [88] survey the state of the art in real-time computing.

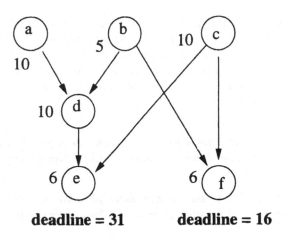

Figure 1.1 A real-time scheduling example from [89]. The nodes represent
the tasks and the directed arcs represent the precedence relation. The number
beside each task is the computation time.

Shin and Ramanathan [89] gave an example shown in Figure 1.1 and Figure 1.2
to show how the scheduling problem in real-time applications is substantially
different from that in non-real-time applications. Figure 1.1 shows a task graph
with two deadlines, assuming that all tasks are ready to execute at time 0 sub-
ject to their precedence constraints. The first deadline indicates that process
e must finish before time 31, and the second deadline requires that process f
must be completed before time 16. The total time to execute all tasks for the
schedule in Figure 1.2(a) is shorter than that in Figure 1.2(b), so the schedule
in Figure 1.2(a) is better for non-real-time-systems, where the performance is
measured by the total execution time to finish all the tasks. However, the deadline 16 to the completion of task f is violated in Figure 1.2(a), whose schedule
is actually useless for the real-time application. As a result, the schedule in
Figure 1.2(b) is what we need for the real-time specification.

Although a problem with a single deadline is a special case of real-time systems,
it is similar to non-real-time problem because both have a single delay value to
minimize. What makes real-time scheduling particularly difficult is the *multiple
deadlines* specification, where interaction among multiple delay values needs to
be considered.

1.3.3   Reactive Systems

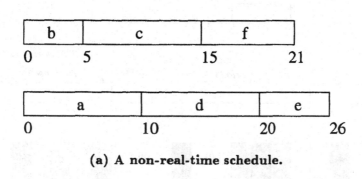

(a) A non-real-time schedule.

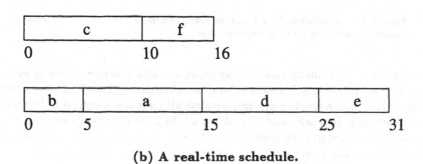

(b) A real-time schedule.

Figure 1.2 Two schedules for the task graph in Figure 1.1. (a) An infeasible schedule with shorter total delay for non-real-time systems. (b) A feasible schedule which satisfies both deadlines for real-time systems.

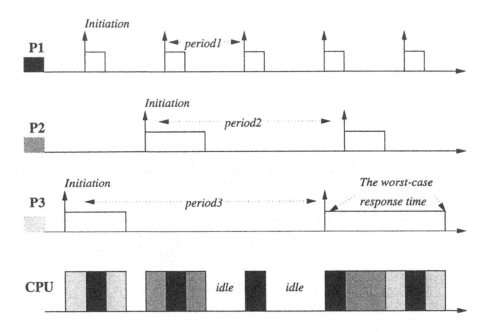

Figure 1.3 A schedule for a real-time reactive system. The idle CPU time cannot be used to reduce the response time.

As a matter of fact, many real-time systems are also reactive systems; sometimes a "real-time system" means a "real-time reactive system". The example in Figure 1.1 is not a reactive system, because all tasks are assumed to be ready at the beginning. In a non-reactive system, all the inputs are ready at the beginning, and all the results are produced at the end; the performance is measured from the beginning to the end. On the other hand, in a reactive system, the interaction with the environment is continuous; the inputs come intermittently and are not all available at the beginning; the outputs are generated in response to the inputs. In a real-time reactive system, there are deadlines for the output response time, rather than for the total execution time. Embedded systems are reactive systems because they interact with the environment directly without human interference.

The rate-monotonic scheduling problem, which will be introduced in Section 2.1.2, is a good example for real-time reactive systems. Figure 1.3 shows a schedule where the idle CPU slots are unable to be filled to improve the performance. In a non-reactive system, all tasks are at our disposal since the beginning, and thus it is straightforward to move the initiation time of processes ahead to fill the idle slots and reduce the future execution time. However, in a reactive system, there is not so much flexibility in scheduling; when a process is ready depends on the environment and is not completely controlled by the scheduler.

1.3.4 Distributed Systems

Many embedded computing systems are distributed systems: communicating processes executing on several CPUs/ASICs connected by communication links [103]. Rosebrugh and Kwang [83] described a pen-based system built from four processors, all of different types. Modern automobiles [94] include tens of microcontrollers ranging in size from 4-bit to 32-bit. Western Digital uses more than one microprocessor and several ASICs in their hard disk drive design.

There are several reasons to build distributed hardware engine for embedded systems [103]:

- Several small 8-bit microcontrollers may be cheaper than a large 32-bit processor, while both implementations can satisfy the performance requirements.

- Multiple microcontrollers may be required to address many independent time-critical devices. A Pentium processor has good overall performance, but when it is busy working on a task, it may not be able to give fast response time to another device at the same time. Another way to look at this point is that according to rate-monotonic analysis [61], when combining many processes into one processor, the processor utilization may become lower and less efficient than distributed systems.

- The devices under control may be physically distributed. A distributed engine can reduce the necessary communication overhead.

1.4 OUTLINE OF BOOK

This book introduces new algorithms for the computer-aided design of real-time distributed embedded systems. The development paths of methodology and dependencies of material in these chapters can be summarized in Figure 1.4.

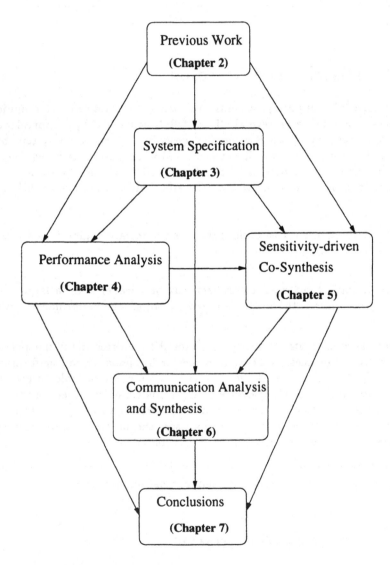

Figure 1.4   The development paths of chapters in this book.

Chapter 2 reviews related work on performance analysis, real-time scheduling, and hardware-software co-design. For performance analysis and scheduling, we discuss work for both uniprocessor and distributed systems. For hardware-software co-design, we review approaches in hardware-software co-specification, hardware-software co-simulation, hardware-software partitioning, and distributed system co-synthesis.

Chapter 3 describes our problem formulation on task graphs and embedded system architecture. Task graphs describe functional and performance requirements of a system, while embedded system architectures model the combined hardware and software implementation. We also propose a communication model for embedded systems.

The performance analysis algorithms are the foundation of co-synthesis. Given a set of task graphs and an embedded system architecture, Chapter 4 presents the performance analysis algorithm with techniques such as fixed-point iteration, phase adjustment, maximum separation, and iterative tightening.

Based on the performance analysis algorithm, given a task graph, a co-synthesis algorithm synthesizes am embedded system architecture. Chapter 5 proposes the co-synthesis algorithm with techniques such as sensitivity analysis, priority prediction, idle-PE elimination, and PE downgrading.

Based on our communication model, Chapter 6 extends the algorithms in Chapter 4 and Chapter 5 to deal with inter-processor communication.

Chapter 7 concludes the contributions of the book and discusses future directions, according to the material proposed in the previous chapters.

2

PREVIOUS WORK

2.1 UNIPROCESSOR SCHEDULING AND PERFORMANCE ANALYSIS

Performance analysis algorithms are critical for system optimization and to ensure that the system meets the real-time deadlines. Software performance analysis is difficult because software execution is less deterministic than the performance of ASICs. The number of clock cycles required for a set of operations on a synchronous ASIC is easy to predict, but the execution time of the same instruction sequence on a CPU may often vary due to cache behavior and interaction with other processes. The performance of an ASIC is seldom affected by adding another ASIC because hardware components run concurrently, while software components cannot be isolated in the same way for a shared processor. The performance analysis of ASICs is considered to be a more well-studied problem [90], so we focus on software performance in this and the next sections.

2.1.1 Single Process Analysis

The performance analysis of a single process has been studied by Mok et al. [69], Puschner and Koza [78], Park and Shaw [73], and Li and Malik [57]. These works calculated the bounds on the running time of *uninterrupted execution* of a sequence of code, and assumed no other processes interfere with the execution of the given program on the same CPU. In most embedded systems, several processes can run concurrently for different tasks and share CPU time according to a certain scheduling criterion. In this case, the *response time*—the time between the invocation and the completion of a process—consists of the whole execution time of the process itself and portions of the execution time of some other processes. Previous work at the process level provides a foundation for the performance analysis at the system level; here, we assume the *computation time*—the uninterrupted execution time of each process—has been given in advance. In order to utilize the results given by those previous works, we need to bear in mind that the computation time for a process is usually not a constant, and should be modeled by a bounded interval for the following reasons:

- The execution paths may vary from time to time in the control flow constructed by conditional branches and while loops.

- The execution time of an instruction is data dependent on some microprocessors. Cache memory in modern architecture causes variation in instruction fetch and memory access time.

- The general problem is difficult because it is equivalent to the halting problem. The performance analysis techniques for a single process often allow some pessimism and cannot guarantee the actual bounds on the running time.

2.1.2   Rate-Monotonic Scheduling Theory

Liu and Layland [61] first studied the scheduling problem of periodic processes with hard deadlines. Suppose each process P_i is characterized by a computation time c_i and a period p_i. A fixed priority is assigned to each process. A process with higher priority can *preempt* or *interrupt* another process, and the CPU always executes the highest-priority ready process to completion. The deadline of a task is equal to its period—i.e. each task must be completed before the next request. In summary, the rate-monotonic scheduling theory has these underlying assumptions:

- All processes run on a single CPU.

- The requests of each deadline constrained process are periodic, with constant interval between requests.

- Deadlines contain run-ability constraints only—i.e., each process must be completed before the next request for it to occur.

- Any nonperiodic tasks in the system, such as initialization or failure-recovery routines, do not have hard, critical deadlines.

- The processes are independent—requests for one process do not depend on the initiation or the completion of requests for another process.

- The uninterrupted computation time for each process is constant and does not vary with time. Context switch time is ignored.

- A process may be initiated at an arbitrary time.

- Each process is given a fixed priority for scheduling.

Liu and Layland showed that the CPU is optimally utilized when tasks are given priorities in order of the tightness of their periods. Such a schedule is called *rate-monotonic priority assignment*. Furthermore, the deadlines of n tasks can be met if the processor utilization U satisfies the bound

$$U \equiv \sum_{j=1}^{i} c_j/p_j \leq n(2^{1/n} - 1) \tag{2.1}$$

When $n = 2$, the utilization is 83%. When n approaches infinity, U converges to 69.3%.

The utilization cannot be 100% because the invocation times of processes are not controllable and the idle CPU time is unable to reduce the peak load when the worst-case performance occurs. The example in Figure 2.1 demonstrates the scenarios. Although the process utilization

$$U \equiv \frac{4}{5} + \frac{7}{14} = 90\%$$

is smaller than 100%, the deadline cannot be met for the second instance of P_2. The idle CPU time slot before the second instance of P_2 is useless in reducing the worst-case response time of P_2, because the system is not allowed to move earlier the request time of the second instance of P_2, which is triggered by external events.

Formula (2.1) is a sufficient condition, but not a necessary condition. In other words, it is possible that the processor utilization exceeds this limit, but the

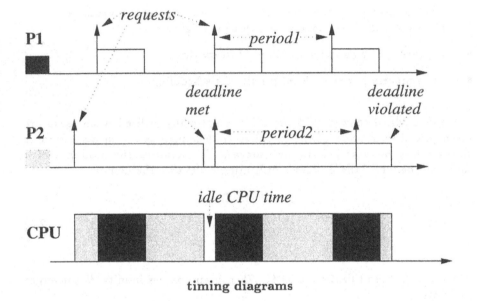

timing diagrams

process	period	computation time	deadline
P_1	10	4	10
P_2	14	7	14

process characteristics

Figure 2.1 Two processes execute under rate-monotonic scheduling. P_1 has higher priority than P_2. The deadline cannot be met for the second instance of P_2.

deadlines are still satisfied. When the processes are not independent and there are precedence relations or release times for them, formula (2.1) becomes more pessimistic.

2.1.3   Extension of Rate-Monotonic Scheduling

The optimality of the rate-monotonic priority assignment and formula (2.1) becomes invalid if the deadline is not equal to the period. When we drive on a highway, the cruise control process in the automobile control may have shorter request period than the anti-lock braking system, but this does not imply that the former is more time critical.

When the deadline is less than or equal to the period, Leung and Whitehead [56] proved that the *inverse-deadline priority assignment* is optimal for one processor. In the inverse-deadline priority assignment, a process with shorter deadline has higher priority.

Lehoczky et al. [53] showed that the worst-case response time w_i of P_i is the smallest nonnegative root of the equation

$$x = g(x) = c_i + \sum_{j=1}^{i-1} c_j \cdot \lceil x/p_j \rceil \qquad (2.2)$$

because $g(w_i) = w_i$ and $g(x) > x$ for $0 \leq x < w_i$. The function $g(x)$ represents the computation time required for processes of higher priority and P_i itself: If the response time is x, there is at most $\lceil x/p_j \rceil$ requests from P_j whose total computation time is $c_j \cdot \lceil x/p_j \rceil$, so $g(x)$ includes these terms for all j as well as the computation time c_i for P_i itself. They enumerated all the multiples of p_j smaller than p_i for $1 \leq j \leq i$ to solve the equation and find values for the w_i's. Later on, the iteration technique to solve the nonlinear equation was mentioned by Sha et al. [88].

Rate monotonic scheduling theory has been generalized in several ways as surveyed by Sha et al. [88]. However, these extensions all apply to uniprocessors.

2.2   SCHEDULING REAL-TIME DISTRIBUTED SYSTEM

A distributed system consists of more than one processor connected by communication links; the processors in the system can have different processing power and different characteristics. There is a large literature on real-time distributed systems. Shin et al. [89] and Ramamritham et al. [81] surveyed this area. Some of this work uses models which are not suitable for embedded systems. For instance, Liu and Shyamasundar [62] and Amon et al. [2] assumed that there can be only one process on each processor; Ramamritham et al. focused only on nonperiodic processes in [82]; Chu et al. [23] applied probablistic approach which is unable to handle hard deadlines. Many scheduling or allocation algorithms for distributed systems focus only on a single nonperiodic task, and cannot handle the RMS model. Chu et al. [22] reviewed works on a single nonperiodic task graph model and categorizes them as graph-theoretic, integer 0-1 programming, and heuristic approaches. However, in most real-time embedded system, different tasks running in different rates mix together. For instance, Chiodo et al. [18] gave a seat-belt example to demonstrate multiple-rate tasks in embedded systems.

In this section, we concentrate on the work which can handle multiple periodic tasks on hard real-time distributed systems.

2.2.1   The LCM Approach

In spite of its wide application, rate monotonic scheduling theory has seldom been extended to schedule distributed systems for periodic processes with data dependencies. Leung and Whitehead [56] showed the analysis and scheduling problem is NP-hard for several cases of scheduling periodic tasks in a distributed system. Many algorithms [36, 80, 76] for periodic tasks in distributed systems form a big task with length of the least common multiple (LCM) of all the periods. Under this approach, the periodic processes are transformed into a big nonperiodic task by duplicating a process for each interval of its period. It is similar to an exhaustive simulation of all possible instances of each periodic process, and the trouble dealing with periodic processes is alleviated. However, we think that the LCM method is impractical for embedded system design for several reasons:

- It is difficult to handle the situation where the period and computation time are bounded but not constant, such as in engines where periods are functions of engine rotation speed. If we use the upper bound of the computation time in the simulation over the length of the LCM, a deadline thought to be satisfied may actually be violated, as shown in the example of Figure 2.2. This phenomenon was also mentioned by Gerber et al. [29] A static table-driven approach [81] to fix all the process invocation times in the length of LCM can handle non-constant computation time, but cannot handle non-constant periods and is inefficient for implementation in embedded systems.

- It is not efficient except some simple cases ($p_1 = 40, p_2 = p_3 = 20$ in [76] or $p_1 = 50, p_2 = 25$ in [80]). In practical embedded systems, the periods can be large (thousands of clock cycles) and coprime. An exhaustive simulation may take more than millions of steps. It becomes worse when we need to simulate all the combinations of possible values in the bounded interval of periods and computation times.

In addition, it also discourages static allocation and scheduling because it treats different instances of a process as different nodes for the length of LCM.

Peng et al. [76] and Hou et al. [36] assumed point-to-point communication with delay proportional to the volume of data. Ramamritham [80] used a single multiple-access network for interprocessor communication.

2.2.2 Analytic Approach for Distributed Periodic Tasks

Leinbaugh and Yamani [55] derived bounds for response times of a set of periodic tasks running on a distributed system without using the LCM enumeration. The system architecture is a network of processors with direct full duplex connection between every pair of PEs. The specification includes multiple tasks with independent periods. Each task consists of segments (processes) with precedence relations between them. The allocation of segments on processors and their computation times are given. Each segment is assigned a fixed priority. Preemptive scheduling is used on processors, while a non-preemptive first-come-first-served (FCFS) schedule is applied on a communication link.

They derived the *guaranteed response times* or the worst-case delays of tasks. If task i has some segments running on the same processor as segments of task j,

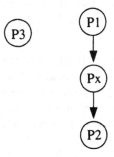

two task graphs

process	period	computation time	allocation	deadline
P_1	100	10	PE1	
P_2	100	10	PE1	
P_3	150	30	PE1	45
P_x	100	[25, 35]	PE2	

process characteristics

Figure 2.2   An example showing that the longest computation time may not give the worst-case response time. The deadline of P_3 is satisfied when P_x runs for 35, but not satisfied if P_x runs in 25. The three processes P_1, P_2 and P_3 share the same CPU and P_3 has lowest priority.

they assumed that the upper bound upon how many times task j can interrupt task i is

$$F(i,j) = \lceil \frac{GT(i) - (MP(j) - GT(j))}{MP(j)} \rceil + 1 \qquad (2.3)$$

where $MP(\cdot)$ is the minimum period and $GT(\cdot)$ is the guaranteed response time. Initially, there is no information about $GT(\cdot)$, so they assumed that $GT(i) = MP(i)$ and $GT(j) = MP(j)$ in the beginning, and formula (2.3) becomes

$$F(i,j) = \lceil \frac{MP(i)}{MP(j)} \rceil$$

Once loose bounds for the $GT(\cdot)$ are computed, they can be used in (2.3) for a more accurate determination of $F(i,j)$. However, formula (2.3) is very pessimistic, because it assumes a process with high priority can preempt a task during the whole interval of the task's execution, even though the task only spends a portion of time on the same CPU the high-priority process is allocated to. Th delay bounds given by their algorithms are not tight enough.

2.2.3 Bus Scheduling

Lehoczky and Sha [54] pointed out that for a distributed system, the scheduling problem has at least two distinct aspects: the scheduling of tasks in an individual processor and the scheduling of message communications among processors across a bus. The problem of processor scheduling has been extensively addressed in the literature, but the problem of bus scheduling with hard real-time deadlines has received little attention. Bus scheduling has a strong similarity to process scheduling: the bus is a resource like a CPU; the tasks are the messages which have hard deadlines.

They argued that there are three important issues which distinguish the bus scheduling problem from the processor scheduling problem:

- *Task preemption*: It is reasonable to assume that a task can be preempted at any time on a processor, but such an assumption is not appropriate for bus scheduling. They solved this problem by assuming that message

periods, lengths and deadlines are integer numbers of slots or packets. A
single packet transmission cannot be preempted.

- *Priority levels*: There may not be enough priority levels for bus scheduling.
  This is a concern for a standard bus or a bus in a general-purpose computer,
  where the number of processors or messages is unpredictable. We believe
  this is not a problem for application-specific systems like embedded system,
  where priority levels can be increased according to the application.

- *Buffering problem*: The deadlines for messages can arise from either task
  deadlines or buffer limitations. Additional buffers may extend a message deadline above the message period, if the task deadline permits. In
  non-real-time system, the buffer size dominates the calculation of message
  deadlines. However, we believe in real-time systems, task deadlines often
  determine the message deadlines and then the required buffer size can be
  inferred from the message deadlines.

They extend rate-monotonic scheduling to handle the bus scheduling problem
with single buffering and sufficient priority levels. Given the maximal period
p, the number of single packet messages n, and the number of distinct periods
m, the worst case bounds on the bus utilization are

$$U = \sum_{i=\lceil (p+1)/2 \rceil}^{p} \frac{1}{i} + \frac{1}{p}(2\lceil \frac{p+1}{2} \rceil - (p+1))$$

$$U = \sum_{i=n}^{2n-1} \frac{1}{i}$$

$$U = m(2^{1/m} - 1)$$

They also extended the results for insufficiency of priority levels and multiple
buffering.

Their results cannot be applied directly to embedded system co-synthesis:

- They did not consider the interaction between the bus scheduling and
  the processor scheduling, especially when the processor cannot perform
  computation and communication in parallel.

■ Their theory is based on rate-monotonic scheduling, and has the same limitations: the deadline must equal to the period; the formulas are sufficient conditions and are not tight enough. Communication messages must receive data from a task and give data to another, so the assumption that they are independent as seen in rate-monotonic scheduling causes undue pessimism.

Sha et al. [87, 86] discussed the design considerations of the IEEE Futurebus+, the standard backplane bus for real-time systems. Futurebus+ provides a powerful connection scheme among large real-time systems. However, its complexity makes the hardware overhead significant. Embedded systems are application specific and need to meet tight cost constraints, so it is not necessary to use a standard backplane bus for real-time scheduling.

2.3 CO-SPECIFICATION AND CO-SIMULATION

This section surveys some co-design frameworks for co-specification and co-simulation. Rowson [84] surveys available simulation techniques for co-design. These systems have not so far provided an *automatic* procedure to assign functions to either hardware components or software running on a CPU. The users need to *manually* decide the system architecture and component to implement each function. Then individual functions may be synthesized on each component chosen by the users. Such systems do provide an integrated environment to manage hardware and software together in co-design projects.

2.3.1 CFSM

Chiodo et al. [18, 19, 20] proposed a unified formal specification model, called a network of Co-Design Finite State Machines (CFSM), for control-dominated systems. The CFSM model is similar to the BFSM model [92]. Both hardware and software use the same CFSM representation. Each CFSM component can be implemented by either hardware or software, but not both. There can be more than one processor for software, but no automated partitioning algorithm was proposed.

Each hardware component is implemented as a fully synchronous Moore FSM, while complex or arithmetic functions are implemented as combinational logic blocks with at most one clock cycle delay. Events are sensed immediately and reacted to within one clock cycle.

Each software component is translated into a block of C code that can be compiled on any processor. There is a private input buffer for each process. All process buffers are checked in turn by a polling or interrupt-based scheduler. Once a process is enabled, it performs at most one CFSM transition.

Seven types of interfaces are designed: hardware to hardware, hardware to interrupt software, software to hardware, software to interrupt software on separate processors, hardware to non-interrupt software, software to non-interrupt software on separate processors, and software to software on the same processor.

2.3.2   Ptolemy

Ptolemy [14], developed at the University of California, Berkeley, is a system-level design framework within which diverse models of computation can coexist and interact. In contrast to the *unified* approach, which seeks a consistent semantics for specification of the complete system, Ptolemy adopts a *heterogeneous* approach, which seeks to systematically combine disjoint semantics. It uses an object-oriented programming methodology in C++ to abstract the differences between models and provide a uniform interface through polymorphism. The Ptolemy system is fundamentally extensible. It defines some base classes in C++ like Star or Domain, and users can create new component models into the system by deriving new classes through inheritance. In summary, Ptolemy is a powerful software system for mixed-mode circuit simulation and rapid prototyping of complex system. Both functional and timing information can be incorporated during simulation.

Hardware-software co-design is a special case of mixed-mode circuit design. Ptolemy can support various system architectures, including multiprocessor systems. A co-design methodology for the synchronous dataflow (SDF) domain [52] in DSP applications has been studied in [43]. Ptolemy can be applied to algorithm-level specification and simulation. After a system configuration is selected, each block is synthesized in either hardware or software separately, and then co-simulation can be used for system verification.

However, users must perform hardware-software partition *manually* and choose the system architecture by themselves. Gupta [30] considers Ptolemy suited for co-simulation but not for co-synthesis. We agree with him on this point because Ptolemy lacks a unified global model to determine how different implementations will affect the overall system performance and cost.

2.3.3 Dual Process Co-Simulation

Thomas et al. [95] proposed a hardware-software codesign methodology developed at Carnegie Mellon University. The system model contains a general purpose CPU, an application-specific hardware, and a system bus. They used two independent communicating processes for co-simulation: the hardware process is performed by the Verilog simulator; the software process is written in a software programming language running in a Unix environment. The communication between these two simulation processes are implemented by means of BSD Unix socket facility and Verilog Programming Language Interface (PLI). The implementation is described in [25]. While timing simulation is one of the methods to evaluate the system performance, their current co-simulator approaches provide only functional simulation ability.

They discuss several algorithmic characteristics of tasks to guide co-synthesis. No automatic algorithm was mentioned under this methodology.

2.3.4 SIERA

Srivastava and Brodersen [90] presented a board-level prototyping framework with hardware-software integration. The application is described by a network of processes. Buffered asynchronous channels are used for inter-process communication. The processes are mapped to a four-layer parameterized architecture template. There are ASICs and CPUs in the architecture. The architecture template has a fixed structure with a certain scalability. The hardware-software partition and the mapping of processes are left to the designers to perform manually.

A simulator is provided for the network of processes in the behavioral level. No co-simulator was mentioned for the architecture template.

2.3.5 SpecCharts

Vahid et al. [97] introduced a new conceptual model, called Program-State Machines (PSM), for the co-specification of embedded systems. They determined five characteristics common to embedded systems: sequential and concurrent behavior decomposition, state transitions, exceptions, sequential algorithms, and behavioral completion. They used SpecCharts, a VHDL extension, to capture the PSM model.

2.3.6 State Chart

Buchenrieder and Christian [13] proposed a prototyping environment for control-oriented hardware-software systems. The system behavior is described as state charts by the Statemate tool. The target architecture consists of a Xilinx XC4000 FPGA, a Motorola 68008 processor, ROM, RAM and I/O ports. Based on the Statemate specification, they proposed a method to generate the hardware subsystem on FPGA, synthesize the interface logic, and use the Statemate code generation to generate C code for the software subsystem.

2.3.7 CASTLE

CASTLE [93] is an interactive environment for embedded systems. The designer manually partitions the algorithmic specification given in C++/VHDL into hardware and software components and refines the architecture step by step. CASTLE quickly provides feedback about the consequences of each partitioning decision. CASTLE can support arbitrary architectures, including not only exchanging one microprocessor type with another type, but also using multiple processors communicating through shared memory.

2.3.8 TOSCA

TOSCA [4, 3] is a framework for control-dominated system codesign. It handles a concurrent model composed of multiple interacting processes represented by state-transition graphs. Basic transformation commands, such as flattening hierarchies and collapsing a set of processes, are provided. Co-simulation of the hardware and software parts is within the same VHDL based environment.

2.3.9   Other On-Going Projects

There are several other on-going research projects on hardware-software code-sign framework, including PARTIF [40], Chinook [21], COBRA [48], COS-MOS [41], and ASAR [7]. We have given a picture on the existing approaches for co-design framework, although we are unable to survey all such projects in detail.

2.4   HARDWARE-SOFTWARE PARTITIONING

In this section, we review some previous work in hardware-software co-synthesis. Unlike the frameworks introduced in Section 2.3, the algorithms described in this section automate the design decision between hardware and software. Most of recent work has studied hardware-software partitioning, which targets a one-CPU-one-ASIC topology with a predefined CPU type. Many co-synthesis algorithms are unable to handle rate constraints and preemptive scheduling, which are common in embedded system design.

2.4.1   VULCAN

Gupta and De Micheli [31, 30] at Stanford University developed the VULCAN co-synthesis system, based on the Olympus high-level synthesis [67]. The input to the co-synthesis system is described in a hardware description language called *HardwareC*, a subset of C defined for hardware description. The input description is compiled into flow graphs which use acyclic graphs to represent control-data dependency. The *non-deterministic* delay or $\mathcal{ND}$ operations are used for variable delay. They introduced *rate constraints* for the reaction rate of each flow graph. The min/max timing constraints can be specified between operations. The target architecture contains a single DLX CPU with one-level memory, hardware components, and a single bus with the CPU as the only bus master.

They construct the software as a set of *concurrent program threads*. They estimate the number of memory accesses, and the software operation delays are the sum of assembly instructions delay, effective address computation time, and memory access time for memory operands. An $\mathcal{ND}$ operation delay is basically the interrupt latency. The operations are then linearized in each thread according to the timing constraints. A first-come-first-served non-preemptive scheduler is used to schedule the threads.

Their software performance estimation algorithm for thread latency can only work on a general purpose register machine with no cache. They assumed that general purpose register machines are the dominant architecture in use today and expected to be so in foreseeable future. This assumption is not true for several reasons: many popular Intel and DSP embedded processors in use today do not belong to this architecture; such an architecture has larger code size and longer interrupt latency due to deep pipeline and plenty of registers to save, so it may not be so prevalent in the embedded system market as in the workstation market.

Their hardware-software partition algorithm proceeds as follows:

- Initially, all operations are implemented in hardware.

- Each operation is considered in the order in the specification. If moving an operation to software does not violate any timing constraints and can reduce the system cost, move from hardware to software at once.

- Whether rate constraints are satisfied is determined by the following two conditions. Let each thread T_i have a rate ρ_i and a latency λ_i, then

$$\frac{1}{\max \rho_i} < \sum_k \lambda_k \qquad (2.4)$$

  Given a bus bandwidth $\bar{B}$, let each variable j transferred between hardware and software have an access rate r_i, then

$$\sum_j r_j < \bar{B} \qquad (2.5)$$

They proposed ideas to extend existing high-level synthesis techniques to co-synthesis problems, especially for linearizing software operations to satisfy min/max timing constraints [32]. The rate constraints are useful for co-design and they pointed out the importance of process-level scheduling to deal with rate constraints. Nevertheless, their methodology has the following restrictions:

- The architecture is limited to a one-CPU-one-bus architecture. The type of the CPU must be a general purpose register machine with one-level memory.

- Although control-flow and hierarchy specifications are allowed, performance constraints are not permitted to cross basic block or hierarchy boundaries. Therefore, the delay analysis is still performed at the basic-block level.

- The software delay analysis algorithm only estimates average performance. This implies that their timing constraints cannot be hard constraints.

- A thread may not be able to start immediately because it needs to wait for the other threads to finish according to the run-time scheduler. They do not estimate such waiting delays, so real-time deadlines cannot be handled.

- The partition algorithm is very greedy. Any move of operation from hardware to software is taken as long as such a move is feasible for satisfying timing constraints. There is no criterion to decide whether one move is better than another and should be taken first.

- Formula (2.5) for bus bandwidth is only a necessary condition instead of a sufficient condition. Therefore, an infeasible communication requirement may be misjudged to be feasible.

2.4.2 COSYMA

COSYMA (COSYnthesis of eMbedded Architectures), developed by Ernst et al. [27, 35] at Technical University of Braunschweig, is an experimental co-synthesis system for embedded controllers. They defined a C^x language for application specification. The application is then modeled by an extended-syntax graph (ES graph), which is a directed acyclic graph describing a sequence of basic blocks. The target architecture contains only one pre-defined microprocessor core and a coprocessor composed of several synthesized hardware modules. The high-level synthesis tool BSS (Braunschweig Synthesis System) [34] was developed to synthesize the co-processor. The communication is implemented by shared memory: the core processor writes to a predefined address, switches to the Hold state, and enables the coprocessor to read.

They adopted a software-oriented approach. Initially, as many functions as possible are implemented in software running on the core processor. External hardware modules are generated only when timing constraints are violated. At present, they use simulated annealing to move nodes in the ES graph from software to hardware. The cost function in simulated annealing is an incremental measure for a basic block to move from software to hardware. The cost increment is calculated according to the timing constraints, the software running time, the hardware performance, and the communication time overhead. They use a hybrid timing analysis algorithm, proposed by Ye et al. [106], to estimate performance. The running time of a basic block is estimated by simulation with the assumption that cache effects are deterministic and the worst case is covered by simulation stimuli. The running time of each basic block is then weighted by its execution count. The total running time for an extended-syntax graph t_s is

$$t_s = t_{SW} + t_{HW} + t_{COM} - t_{HW/SW}$$

where t_{SW} is the total running time of software, t_{SW} is total running time of hardware, t_{COM} is the communication time, and $t_{HW/SW}$ is the overlapping time when software and hardware are both running. They set $t_{HW/SW}$ to zero because software and hardware must run in an interleaved manner in their architecture.

The system is an experimental platform for hardware-software partition algorithms and not restricted to a particular method. The C^x language provides a flexible and powerful programming capability for the users. However, their current approach has some limitations:

- The architecture is limited to a one-CPU-one-ASIC architecture. The type of the CPU cannot be changed during synthesis.

- The software components and hardware components are not allowed to run concurrently in their model. This limits the resource utilization.

- Their performance analysis algorithm cannot handle periodic and concurrent tasks. The simulation-based approach may not be accurate for hard timing constraints.

- The cost function in simulated annealing does not account for hardware costs. The user can only specify a limit on the number of functional units in the coprocessor. The cost of each individual hardware unit is not taken into consideration during hardware-software partition.

2.4.3   Binary Constraint Search

Vahid et al. [96] proposed a binary-constraint search algorithm for hardware-software partition. The algorithm is based on an iterative improvement partition algorithm such as simulated annealing. Given the observation that decreasing performance violations usually increases hardware size and *vice versa*, resulting in large "hills" to climb, they relax the cost function goal by artificially setting a larger hardware size constraint. Initially, the hardware size constraint is in a range from zero to the size implementing all the functions in hardware. The hardware size constraint is set to the middle of the range for simulated annealing. If the result of simulated annealing satisfies the hardware size constraint, the range is updated to the upper half of the current range; otherwise, it is updated to the lower half of the current range. Another simulated annealing is started for the new range, and the similar steps are repeated until the range is reduced to a certain precision factor.

They provided a new method to improve pure simulated annealing approach and reduce the chance to stop at a local minimum during hardware-software partition. Nevertheless, there are a few restrictions in their approach:

- The system architecture is restricted to a one-CPU-one-ASIC architecture. The type of the CPU must be predefined.

- It cannot handle rate constraints.

- They do not mention how to schedule processes in software during synthesis.

2.4.4 UNITY Language Partition

Barros et al. [8] used a multi-stage clustering technique to partition a system specified by the UNITY language. The UNITY language is not as powerful as popular languages like C or VHDL; it consists of only assignment statements. The target architecture contains only one CPU and several hardware functional units. The cluster tree is built from the statements according to the distance information like parallelism degree, data dependencies, and resource sharing. A cut line is then chosen at a certain level of the tree to divide the statements into clusters. Each cluster will be allocated to either the CPU or functional unit. The allocation is decided mainly by the type of the assignments in the UNITY language instead of delay or cost information. Their clustering algorithm helps increase resource sharing in functional units, but can be applied only to a very restricted model specified by the UNITY language.

2.4.5   C Program Partitioning

Jantsch et al. [42] added a separate phase into the GNU C compiler to partition
the implementation of a C program on a Sparc CPU and an FPGA chip. The
hardware-software partitioning algorithm generates assembly code for the software parts and behavioral VHDL code for hardware parts. They also mapped
the virtual CPU registers allocated by the GNU C compiler onto FPGA registers to minimize memory interface traffic and improve performance. A few
regions in the C programs are selected as candidates for hardware implementation. With each candidate a gate count and a speedup factor are associated.
Given a gate count limit on the FPGA, the partitioning problem is formulated
as a knapsack problem. A dynamic programming technique is used to solve the
knapsack problem.

Their approach is very useful to speedup a C program by considering an FPGA
as a hardware accelerator. For embedded system design, the approach is not
very general:

- The architecture must be a one-CPU-one-FPGA architecture.

- The software and hardware are not allowed to run concurrently.

- They did not mention how to estimate the gate count and speedup for
  implementing a software part in hardware.

- The total run time of a single C program is the optimization goal. Rate
  constraints or multiple real-time deadlines are not handled.

2.4.6   Design Assistant

Kalavade and Lee [44] proposed a GCLP algorithm for hardware-software partition. The application is represented by a directed acyclic graph, where nodes
are processes and arcs are data or control precedences between nodes. Associated with each node are four non-negative numbers: area required for hardware implementation, code size for software implementation, execution time
for hardware implementation, and execution time for software implementation.
Associated with each arc is the number of samples for data communication.
An area and time for a data sample communication are given. There are three
design constraints: the latency, the hardware capacity, and the software capacity (available memory). The GCLP algorithm will determine the mapping
(hardware or software) and the start time for each node.

Unlike VULCAN or COSYMA, which use *transformational* methods starting from an initial solution, GCLP is a *constructive* approach which traverses the graph and allocates nodes step by step. Similar to *list scheduling* in high-level synthesis [65], some measures are necessary to set a priority for ready nodes to choose. The *global criticality* (GC) is a measure of time criticality at each step of the algorithm, based on the latency requirements. The *local phase* (LP) is a measure of node heterogeneity characteristics, based on the implementation preference and area/time gain on different implementation. Instead of a hardwired objective function, the algorithm uses GP and LP to dynamically select an appropriate objective between area and time at each step, and switch nodes between hardware and software.

The GCLP algorithm has been integrated into Design Assistant [45], a system-level codesign framework. The Design Assistant consists of tools for partitioning, synthesis, and co-simulation.

The GCLP algorithm is efficient, and takes both delay and area into consideration during synthesis. However, it has the limitations listed below:

- The system architecture is restricted to a one-CPU architecture. The type of the CPU must be predefined.

- There is only one task. Neither rate constraints nor preemptive scheduling is considered.

- Processes implemented in hardware do not run concurrently, and are scheduled in the same way as those in software.

- Each communication data sample has dedicated hardware for it. Communication hardware cannot be shared by different arcs even though they are scheduled in disjoint slots.

2.5 DISTRIBUTED SYSTEM CO-SYNTHESIS

Distibuted system co-synthesis should automate not only the mapping of tasks into CPUs and ASIC, but also the selection of the hardware engine topology. There can be multiple processors. The co-synthesis algorithms need to select the number and types of processors and the number and types of communication links, in addition to allocating and scheduling of processes on processors.

2.5.1   SOS

Prakash and Parker [77] formulated heterogeneous multiprocessor system synthesis as a mixed integer linear program (MILP) in SOS, a formal synthesis approach developed at University of Southern California. The problem is specified by the *task model* and the *system model*. In the task model, the application is specified in terms of a single task graph, which is a directed acyclic data flow graph. A node in the task graph is a subtask (process). The system model describes the architecture, where there can be several types of processors with different cost-speed characteristics. The communication topology is restricted to either point-to-point interconnection or a single system bus. The interprocessor communication delay is fixed. There are 0-1 variables for each pair of subtask and processor to represent the subtask-to-processor mapping, and for each data arc in the task graph to represent remote/local data transfer. There are also timing variables for the available time, start time, and end time of each subtask or data transfer. The various linear or nonlinear constraints among these variables can be converted into an MILP formulation. Given the cost and the computation times of subtasks on each type of processor, they could simultaneously allocate and schedule the subtasks while designing the underlying distributed system. They handle both communication delay and communication cost during co-synthesis.

However, such a formal approach has the following limitations:

- It cannot deal with more than one task graph concurrently running under different initiation rates and real-time deadlines.

- The MILP formulation cannot handle preemptive scheduling of subtasks.

- The mathematical programming methodology is optimal but not efficient for large examples. For instance, their MILP algorithm required hours to execute for a task graph composed of only 9 processes.

- The inter-processor communication topology is fixed in advance, rather than synthesized. It is assumed that there is a dedicated communication hardware for each processor such that computation and communication can be performed in parallel.

2.5.2   Architectural Co-Synthesis

Based on Prakash and Parker's formulation, Wolf [102] developed a heuristic algorithm for architectural co-synthesis of distributed embedded computing systems. In addition to it, the 69.3% processor utilization limit derived from formula (2.1) in rate-monotonic analysis is used to approximate the scheduling feasibility for rate constraints. The results of the heuristic algorithm are comparable to the MILP formulation in many cases, but the algorithm is much more efficient than solving MILP.

2.5.3   Optimal Part Selection

Haworth et al. [33] proposed an algorithm that chooses parts from a part library (or catalog) to implement a set of functions and meet cost bounds. The problem becomes difficult when there are multiple-function parts. Each part has a *cost-attribute-scores vector* that represents costs along various attributes. The total cost for each attribute is assumed to be computed by an additive, monotonic function. There is a bound for each attribute. The algorithm selects a set of parts which is *nondominated*—there does not exist another part set whose cost-attribute score is better for each cost-attribute.

They construct a *part-function-attribute table*, and apply approaches similar to the prime-implicant-chart-reduction algorithm in the Quine-McCluskey method for minimizing switching functions. The worst-case performance of their algorithm is exponential in time and space, but they use some techniques to offer substantial performance improvement.

However, their part selection algorithm does not consider timing properties at all—timing attributes cannot be expressed simply by additive and monotonic functions. No performance or scheduling issues are discussed. They dealt with only hardware parts—the functions on the same part can run concurrently without interfering with one another. The communication overhead is ignored.

2.5.4   Configuration Level Partition

D'Ambrosio and Hu [26] presented a hardware-software co-synthesis approach at the configuration level, where hardware is modeled as resources with no detailed functionality and software is modeled as tasks utilizing the resources. Their approach determine the overall system architecture, including the number of processors, the allocation of functions to either software or hardware components, and scheduling the functions implemented in software.

The processes have different initiation periods, computation times, release times, and deadlines. Their algorithm utilizes the optimal part selection algorithm. In order to take timing property into consideration during optimal part selection, a *feasibility factor* is defined to give a rough measure about whether a given system configuration is feasible to satisfy the deadlines. The rate monotonic priority assignment is adopted for scheduling—the smaller the period, the higher the priority. The feasibility factor is calculated based on formula (2.1) in rate monotonic scheduling theory.

Because the feasibility factor is not an accurate performance measure, the feasibility factor is considered as an attribute in optimal part selection, and the optimal part selection gives a Pareto-optimal set of solutions. Each solution in the Pareto-optimal set implements all the functions, satisfies all the constraints, and there is no other solution with all the attribute values better than that of the solution. Then the solutions are sorted by the total costs, and screened for feasibility by a TASSIM event-driven simulator [37] from the least cost solution to the highest. The first feasible solution encountered is chosen as the final solution.

Their approach can co-synthesize the underlying system architecture, and handle rate constraints and real-time deadlines. Nonetheless, the approach has the following restrictions:

- The complexity in enumerating all possible solutions in the Pareto-optimal set restricted them to explore only one-CPU system configuration.

- The scheduling algorithm is based on rate monotonic scheduling theory, so no data dependencies between processes are considered.

- The final performance analysis relies on simulation. Simulation is both time-consuming and not guaranteed to prove feasibility.

- Communication overhead is ignored.

2.6   EVENT SEPARATION ALGORITHMS

Methodologies in research areas other than hardware-software co-design may be applied to the co-synthesis problem. Although the algorithms which derive maximum event separations are not directly related to the hardware-software co-design problem, since we will apply a maximum separation algorithm to tighten performance estimates in Section 4.4, this section reviews previous work on such algorithms.

The situation where the delay is specified with a lower bound and an upper bound instead of a precise constant has been encountered in other research areas, such as interface timing verification [28, 66, 98, 108], asynchronous circuit synthesis [70, 51, 2], and optimal clock scheduling for level-sensitive latch circuits [85]. Such problems are often modeled by an *event graph*. In an event graph, a delay constraint c_{xy} with a lower bound $c_{xy}.lower$ and an upper bound $c_{xy}.upper$, belongs to one of the three types:

- **linear:** For all x such that c_{xy} is linear,

$$\max_x(x.time + c_{xy}.lower) \leq y.time \leq \min_x(x.time + c_{xy}.upper)$$

- **max:** For all x such that c_{xy} is max,

$$\max_x(x.time + c_{xy}.lower) \leq y.time \leq \max_x(x.time + c_{xy}.upper)$$

- **min:** For all x such that c_{xy} is min,

$$\min_x(x.time + c_{xy}.lower) \leq y.time \leq \min_x(x.time + c_{xy}.upper)$$

An example in Figure 2.3 shows how different kinds of constraints can generate different event times given the same bound values in the constraints. In Figure 2.3(b), $z2.time = 7$, which does not satisfy the upper bound 3 in the max constraint from $y2$ to $z2$. In Figure 2.3(c), it is allowed that $z3.time = 1$, which does not satisfy the lower bound 2 in the min constraint from $y3$ to $z3$.

It is important not to confuse the max and min constraints described here with the maximum or minimum constraints used in some other literature [9, 49]. Generally, those maximum or minimum constraints are the same as the upper bounds or lower bounds in the linear constraints. While *all* the lower bounds and *all* the upper bounds of linear constraints must be satisfied, only *one* upper bound of the max constraints entering a node has to be satisfied and only *one* lower bound of the min constraints entering a node has to be satisfied. This nonlinearity of max and min operators makes the problem more complicated than the purely linear constraint cases such as layout compaction [60, 105], which can be solved by longest path algorithms.

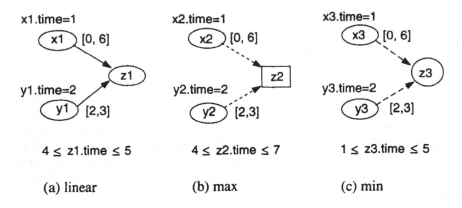

Figure 2.3 Suppose the times of the events x's and y's are fixed. The range of the occurring times of event z's can be different according to the type of the constraints.

Gahlinger [28] pointed out all the three types of constraints are necessary for modeling interface timing and described how maximum event separations are used to check the satisfaction of timing requirements. Vanbekbergen et al. [98] gave a graphic interpretation of different types of delay constraints. Their type 1, type 2, and type 3 constraints are the same as linear, max, and min constraints respectively.

The event separation problem with only linear constraints was studied by Borriello [10] and by Brzozowski et al. [12], who use shortest-paths algorithms to solve the constraint graph. Vanbekbergen et al. [98] proposed an $O(V^2)$ algorithm, and later McMillian and Dill [66] presented an $O(E)$ algorithm, for the acyclic max-only constraint problem.

So far no algorithm with a proven polynomial-time bound has been published for any mixed-constraint problem. McMillian and Dill [66] gave an iterative tightening algorithm for the problem with max and linear constraints. They start with separation values greater than or equal to maximum separations and decrease the separation values to the achievable maximum separations step by step. The running time is bounded by $O(V^3 S)$ where S is the sum of all the bound values on the constraints. As a result, the running time depends on delay values and might not have a finite bound for convergence when the constraint bound values can be infinite. There are examples for which their algorithm has exponential running time.

Walkup and Borriello [100] propose a *short circuiting* algorithm for max-linear constraint problems. They use strongly connected component analysis on the *dependency graph*, a subgraph constructed by constraint updating, to discover possible negative cycles and avoid going around the negative cycles many times. The convergence of the short circuiting algorithm will not be affected by the existence of infinite bounds on the constraints. However, in order to guarantee the convergence, it needs the assumption that no event separation can be infinite.

McMillan and Dill [66] prove that the problem with both max and min constraints is NP-complete. They solve the problem with all the three types of constraints by eliminating min constraints and dividing the problem into subproblems which contain only linear and max constraints. Since min constraints are few in most applications, the number of subproblems may not be very large in practice.

Burks and Sakallah [16] apply mathematical programming techniques to solve the general min-max linear programming problem. Although their approach can be used to solve the timing problem, by concentrating on only the special case of the general min-max linear programming problem, we are able to develop a much more efficient graph-based algorithm.

Yen et al. [107, 108] present an efficient algorithm for computing separations between events under the min-max linear constraint model. It uses two auxiliary graphs—compulsory constraints and slacks—to quickly check constraint satisfaction. Unlike previous algorithms, the algorithm must converge to correct maximum separation values in a finite number of steps, or report an inconsistency of the constraints, irrespective of the existence of infinite constraint bounds or infinite event separations. It is conjectured to run in $O(VE + V^2 \log V)$ time, when only linear and max constraints exist.

3

SYSTEM SPECIFICATION

3.1 TASK GRAPH MODEL

We describe the embedded system behavior by *task graphs*. The task graph model is similar to those used in distributed system scheduling and allocation problems [36, 80, 76, 55]. The task graphs represent what functions to perform and their performance requirements, but are not related to how these functions are implemented, either in hardware or software.

3.1.1 Processes

A **process** is a single thread of execution. Some processes may be implementable by either a CPU or an ASIC; we assume that the processes have been partitioned so that they do not cross CPU-ASIC or CPU-CPU boundaries. The designer can decide the granularity of the partition and manually divide an application into processes when using high-level programming languages. A partition with a fine granularity provides more flexibility for scheduling and allocation during co-synthesis, and reduce the run time for compilers and hardware synthesis tools which synthesize each process; but a partition with a coarse granularity reduces inter-process communication overhead, increases the optimization opportunity in compilation or synthesis of a single process, and reduce the run-time of the hardware-software co-synthesis algorithm. This work does not discuss automatic partition methodology.

A process is characterized by a **computation time**, which is the uninterrupted execution time of the process. The computation time may not be a constant but should be bounded for real-time applications. The computation time of

PE type	Computation time					
	P_1	P_2	P_3	P_4	P_5	P_6
X	179	100	95	213	367	75
Y	204	173	124	372	394	84
Z	-	-	30	-	-	-

Table 3.1 An example table showing the computation time of process $P_1, \cdots, P_6$. The computation time is a function of PE type.

a process P_i is $[c_i^{lower}, c_i^{upper}]$ where c_i^{lower} is the lower bound, and c_i^{upper} is the upper bound. As mentioned in Section 2.1.1, a computation time needs to be modeled by a bounded interval due to factors such as conditional behavior in the control flow, cache access, and the inaccuracy of process-level analysis techniques. Unless specified otherwise, we will often use c_i to represent c_i^{upper} to simplify the notation. Without loss of generality, we assume that a time unit is chosen such that all the numbers for time are integer. If a computation time cannot be represented accurately by an integral multiple of the time unit, it can still be bounded by integers. For example, if $c_1 = 627.4$, it can be described by an interval [627, 628]. In practical systems, infinite precision is impossible for all numeric values, so such approximation is unavoidable.

The computation time is a function of the PE type to which a process is allocated. We often use a table to show the computation time of a process on each type of PE which can implement the process, as shown in Table 3.1. If a certain PE cannot implement a particular process, the corresponding entry in the table is empty or can be artificially set to infinity. For instance, in Table 3.1, if PE type Z is an ASIC which is designed to implement process P_3 only, the computation times for other processes on Z can be set to infinity.

In most embedded systems, several processes can run concurrently for different tasks and share CPU time according to a certain scheduling criterion. In this case, the **response time**—the time between the invocation and the completion of a process—will consists of the whole computation time of the process itself and portions of the computation time of some other processes.

Starting the problem model at the process level instead of discussing an specialized co-design model has the following advantages:

■ It is general enough to be compatible with most other models. As pointed out in Ernst et al.'s comments [27] on Gupta et al.'s work [31], using pure hardware model for specification can introduce limitations in describing complex software constructs like dynamic data structure, while software development is actually an important problem in embedded system design. A process with detailed min/max interface timing constraints is likely to be implemented only by hardware. Interface timing constraints are often in the range of a few clock cycles and software instructions are unable to accurately control such low-level timing, because interrupt latency, context switch time, cache misses in instruction fetches are much larger than this range. On the other hand, a process with dynamic data structures, pointers, or recursive function calls tends to be implemented in software, since no industrial synthesis tools nowadays can support these features. A process with a table showing which PEs (hardware or software) can implement it may be either of the above two extreme cases. The notion of processes is compatible with most software programming languages like C++/C or Pascal, and hardware design languages like VHDL or Verilog. Focusing on the process level avoids the necessity to define a new co-design language before related research work has matured.

■ It allows us to utilize related research. Previous work on embedded software performance analysis [57], for example, can be applied to get the computation time of processes implemented in software. Such on-going research provides stronger capabilities, such as cache effect analysis [59], than ad hoc approaches for co-design [27, 30]. Similarly, code generation techniques for embedded processors [6, 91] and high-level synthesis techniques [65] can be used to synthesize processes on PEs. It is better for co-synthesis algorithms to be constructed based on other research results than to start from scratch.

■ It provides more flexibility in the granularity of partitioning. The size of a process can be controlled by the designer or an automatic partition algorithm. Sticking to the operator level, as done in high-level synthesis, may cause unnecessary hardware-software communication overhead and inefficiency of co-synthesis algorithms.

3.1.2 Tasks

A **task** is a partially-ordered set of processes, which may be represented as an acyclic directed graph known as a **task graph**. A directed edge from process P_i to process P_j represents a data dependency: the output of P_i is the input of

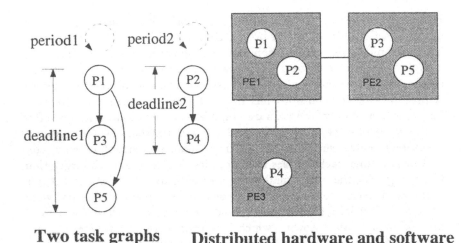

Two task graphs   Distributed hardware and software

Figure 3.1 An embedded system specification and implementation.
$P_1, \cdots, P_5$ are processes.

P_j; P_j needs to wait for P_i to finish in order to start. We assume that a process
will not be initiated until all its inputs have arrived, that it issues its outputs
when it terminates. A problem specification may contain several concurrently
running tasks. An embedded system may perform several tasks which are
nearly decoupled, such as running several relatively independent peripherals;
in such cases, it may be useful to describe the system as unconnected subsets of
processes. Processes and tasks are illustrated on the left-hand side of Figure 3.1.
We introduce a START node and an END node for each task. The START
node represents invocation time instant for running the task. The END node
stands for time instant at the completion of all the processes in one execution
of the task. When the START node is initiated, all the processes without
predecessors in the task graph are invoked. The END node is reached when
all processes in the task finishes their execution. Define the **worst-case task
delay** or **worst-case response time** as the longest possible elapsed time from
START to END for each execution of the task. Each task is given a **period**
(sometimes referred to as a **rate constraint**) which defines the time between
two consecutive invocations of the task, and a **hard deadline** which defines
the maximum time allowed from invocation to completion of the task, and a
soft deadline which describes the optimization goal of the task delay. If a
task τ_j is not issued in a constant rate, the period is modeled by an interval
$[p_j^{lower}, p_j^{upper}]$. Unless specified otherwise, we will often use p_j for p_j^{lower} to

simplify the notation. The periods may be bounded intervals for the following reasons:

- The invocation of tasks is often triggered by external events. The rate of events in a real-world environment may not be constant.

- Rates specified to be identical may not be synchronized. As an example, if two modems receive data at 2400 bits/second, the incoming data rates cannot be considered as the same. If we know the deviation from the specification is within 1 percent, the synchronization problem can be relieved by using a bounded rate in the interval of (2397, 2403).

A deadline can be satisfied if it is greater than or equal to the worst-case task delay. A hard deadline must be satisfied; missing a hard deadline will cause system failure. Meeting a soft deadline is desirable, but failure to meet it is acceptable for correct system function.

It is assumed that the period of a task must be larger than the computation time of each process in the task. If the computation time of a process is larger than the task period, no matter whether the process is implemented by hardware or software, it may not be able to finish before it starts next time. The process can nonetheless be partitioned into smaller process such that the computation time becomes smaller than the task period.

A **dummy process** is a process with zero delay, or a process with nonzero delay but not allocated to any physical PE. Dummy processes do not consume any resources but can provide extra modeling capability to task graphs. A process in a task may have a *release time*—the process cannot be initiated until a certain time after the invocation of the task—due to late-arriving inputs. A task can have several deadlines to the completion of different processes because of different performance requirements. The release times and multiple deadlines can be modeled by inserting dummy processes in the task graph, as described in Figure 3.2. Therefore, without loss of generality, we assume each task has a single invocation time instant (the single START node) and a single deadline.

3.1.3   Data Communication

In a task graph, a weight on a process denotes the volume of output data for communication through all data dependency edges emanating from the process. Without loss of generality, we assume that the data transferred on each data

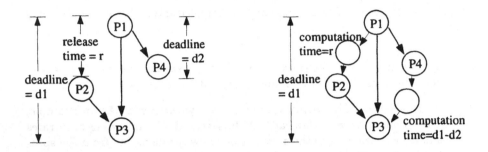

Figure 3.2 A task graph with a release time and two deadlines (left-hand side) can be transformed into a task graph with a single deadline (right-hand side) by adding two dummy processes (represented by empty circles).

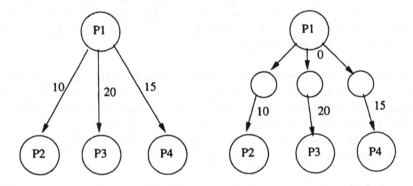

Figure 3.3 If the communication data output from process P_1 are different for each edge (left-hand side), dummy processes can be inserted such that the output data are the same for each outward dependency edge from a process (right-hand side). These dummy processes need to be allocated to the same PE as P_1.

dependency edge emanating from a process are the same. If the data sent on different edges are different, we can create a dummy process between each edge and the process sending data, as shown in Figure 3.3. These dummy processes have zero delay and must be allocated to the same PE as the process sending data. Each of the dummy processes receives zero data from the sending process and sends data of volume specified by the weight below it to the outward data dependency edge.

Because two processes on an edge runs in the same task rate, each data item sent out must be consumed once by the receiving process. In other words, the send and the receive action must be performed in the same rate (synchronized). An **inter-task communication** is a data transfer between processes in two different tasks. Because tasks run at different rates, unlike intra-task data dependencies, such communication does not have to be synchronized. An example for inter-task communication is a status variable whose most recent data are sampled to show the current status; several data sent between the samplings may be discarded; the same data sent can be used several times if a new data item has not arrived. Another example can be found in the case where several transferred data are consumed together in a batch, as seen in communication protocol and image processing. The cost and run time for inter-task communication affect the performance of other processes and cannot be ignored.

3.1.4   Limitation of Task Graphs

Our current task graph model does not handle some system behaviors very efficiently:

- It does not explicitly represent control flow information like conditional modes and loops in the task level, although such control constructs can exist arbitrarily inside any single process.

- It does not handle hierarchy representation, as seen in function calls or component instantiation in high-level languages.

Nevertheless, we may work around these limitations as follows:

- Control and data can actually be formulated in the same model. For conditional branches, the task graph can still represent them like a control-data-flow graph (CDFG). We may get a pessimistic results by this approach, because the exclusive relation between processes in two exclusive conditional paths is not utilized. For instance, suppose processes P_1 and P_2 are in exclusive conditional paths, and both of them have higher priority than P_3; then either P_1 or P_2, but not both, can preempt P_3 at a time. However, the task graph will allow both P_1 and P_2 to preempt P_3 at a time for the worst case.

- For conditional branches, we can also enumerate all possible execution paths or running modes, and handle them separately. Since the number of execution paths may grow exponentially with the number of nodes, the running time of the co-synthesis algorithm may become too long.

■ If there is a bound on the number of iterations for a loop, the loop can be unrolled into a normal task graph, although it is not efficient to do so. If there is no bound on the number of iterations for a loop, the task cannot have real-time deadlines, and its processes can be allocated to cheap PEs with lowest scheduling priorities.

■ For hierarchical specification, the hierarchy can be flattened into one-level hierarchy which can be represented by a task graph. Flattening may cause inefficiency in both the run time of the co-synthesis algorithm and embedded system implementation.

The above approaches may not be efficient. Nonetheless, this work provides groundwork for future enhancements. In the scheduling problems of high-level synthesis, a huge amount of early work focused on only datapath scheduling, which has similar limitations. Several articles provide good surveys of datapath scheduling algorithms [75, 65, 39]. Later on, new approaches for scheduling control-dominated designs were proposed to handle a more general model [17, 115, 109, 114]. We expect that the on-going research will provide similar improvements for hardware-software co-synthesis in the future.

3.2 EMBEDDED SYSTEM ARCHITECTURE

Given the functions specified by task graphs, our goal is to simultaneously design the **hardware engine** which consists of a network of heterogeneous **processing elements** (PEs), either CPUs or ASICs, and the **application software architecture** which implements the required functions. The application software architecture consists of allocating functions to PEs in the hardware engine and scheduling their execution. We refer to the hardware engine and application software architectures together as the **embedded system architecture**, since both the hardware and software architectures contribute to the system design.

The embedded system architecture describes the implementation scheme for the application. For the co-synthesis algorithm, while task graphs are inputs, an embedded system architecture is the output of the algorithm. For the performance analysis algorithm, both task graphs and the embedded system architecture are inputs of the algorithm.

3.2.1 Hardware Engine

As depicted in the right-hand side of Figure 3.1, the hardware engine is a labeled graph whose nodes represent **processing elements** (PEs) and whose edges represent **communication links**.

The co-synthesis algorithm requires the designer to provide a list of available **PE types**. A PE type can denote an off-the-shelf microprocessor, a catalog ASIC, or a custom ASIC. Each PE is associated with a certain PE type. The set of PEs in the hardware engine is not equivalent to the set of available PE types—it is not necessary that every PE type is used in the hardware engine; there may be more than one PE of the same PE type. We assume that there exists a method which generates the performance measure of a process for each PE type. Such a method can be a process-level software analysis algorithm, or a timing analysis algorithm for hardware synthesis. The computation times of processes are given by these methods.

There is a cost associated with each PE type. Cost can be the price, area, or power consumption of the system. A cost includes the cost of necessary peripheral chips like memory, I/O ports, and a clock generator. A cost can be a function of process allocation. For instance, when one more process is allocated to a CPU, the memory or I/O port requirements may or may not increase. Such increases are often nonlinear functions.

The delay and cost of interprocessor communication plays an important role in distributed embedded systems. Most previous work simplify the communication model by assuming point-to-point communication or a single-bus architecture, which may not meet practical design methodology. Our communication model will be elaborated in Section 3.3.

We assume that the total **system cost** is the sum of all the component costs in the hardware engine. There are a **hard cost constraint**, and a **soft cost constraint** on the total system cost. The hard cost constraint cannot be violated. The soft cost constraint is an optimization goal to achieve but not a requirement for system feasibility.

3.2.2   Allocation

The **allocation** of processes is given by a mapping of processes onto PEs. We assume *static allocation* of processes—every instance of a periodic process is executed on the same PE; the allocation will not change during execution. In a heterogeneous system, different CPUs have different instruction sets. In an embedded system, program codes are stored statically on ROMs, rather than dynamically loaded from disk drives or networks. It is very inefficient to put several copies of the same program in the memory of different PEs. Changing allocation during run time may also induce run time overhead, duplication of I/O ports and communication links for the same process on different PEs.

The designer can specify **process cohesion** or **process exclusion** relations for any pair of processes. Process cohesion means these two processes must be allocated to the same PE. If two processes share the same external signal set, a resource, a large amount of data, or the same functionality which is too complicated or too expensive to duplicate, the designer may want to force the co-synthesis algorithm to allocate them on the same PE. The dummy processes in Figure 3.3 is an example of process cohesion. Process exclusion means two processes must be allocated to different PEs. If two processes interface two set of time-critical signals, and these two set of signals are physically separated (10 feet apart), the designer can let them be implemented on two different PEs.

3.2.3   Scheduling

It is possible to allocate more than one process to the same CPU; in this case, we need to schedule their execution because only one process can run on a CPU at a time. Scheduling is a widely-researched topic. Scheduling tasks has several possible interpretations in different research areas. For instance,

- When a compiler schedules instructions during code generation, it gives the order of instructions without knowing the exact starting time of each instruction.

- In high-level synthesis or job-shop problems in operation research, a scheduling algorithm explicitly specifies a absolute or relative time instant for each operation to start.

- When there is no fixed order or fixed arrival time for processes which share a CPU, a scheduling algorithm statically or dynamically assigns a priority to each process. The CPU always executes the highest-priority ready process.

When the requests of tasks are periodic and the periods are different, it is difficult to give a strict order of tasks and it is inefficient to list the starting times for all instances of each task. When there is variation in process computation times or task periods, it becomes impossible to schedule the starting time in advance. Priority scheduling is perhaps the most practical approach for embedded systems.

Consequently, the **schedule** of processes is an assignment of priorities to processes. Each process is given an integral priority: no two processes allocated to the same CPU have the same priority and the CPU always executes the highest-priority ready process to completion. Unless mentioned otherwise, process execution is *preemptable*—it can be interrupted by another process with a higher priority. The operating system overhead such as interrupt latency, context-switch latency is not negligible, although with fixed priority scheduling, operating system overhead is frequently limited. However, the effects of such performance overhead can be straightforwardly incorporated into the computation time of a process.

3.3 THE BUS MODEL

Buses are widely used in many practical embedded systems. Without loss of generality, a point-to-point communication link can be modeled as a bus connecting only two PEs.

In the computer industry, there are many standard buses [63] which allow I/O devices manufactured by different companies to be plugged into the same computer system. Most of these standard buses belong to *backplane* or *I/O* buses [74]. By a modular connection scheme, a computer system can be extended by adding new devices which were unknown during the design phase of the system. However, we do not specifically assume the use of any standard bus inside an embedded system for interprocessor communication, though the embedded system may contain functions which interface with some other systems through a standard bus. The standard buses require more overhead in both hardware and performance in exchange for the advantages of scalability or extensibility, which is not necessary for an embedded system, because embedded systems are application specific. For instance, in both FutureBus+ [87] and small computer system interface (SCSI), a sophisticated controller is required for interfacing each component including memory. The cost of a controller may be close to a small PE. In embedded systems, the cost of these controllers can be replaced with extra PEs or dedicated buses for processes with tight deadlines to increase performance.

3.3.1   Shared Memory Communication

We assume each CPU has a local memory where the program code and local data are stored. In other words, the instruction fetch and the access of local data on the local bus will not be affected by the activities on the buses for interprocessor communication. In this work, unless specified otherwise, the word *bus* is used to mean only those buses which connect multiple PEs, and we will not analyze or synthesize local buses connecting a CPU with its local memory and I/O ports. A local bus is associated with the CPU and its cost has been incorporated into the corresponding PE cost. The data transfer between two processes allocated to the same PE incurs no extra delay or cost, because one process can output the data directly on the local memory location where the other process keeps its input data. On the other hand, when two processes are allocated to different PEs, the communication between them needs to go through a bus and can introduce a delay in addition to the execution of the processes.

A CPU can have a portion of address space shared with another PE on a bus. A data item transmitted from a PE to another will be put in a shared memory location, which may be a simple buffer when total data transferred have only one or two bytes. Such a bus can be classified as a *CPU-memory bus* [74], where processor and memory can be connected directly to the bus without a special controller. Since a data item can be identified *implicitly* by its memory location, there is no need for a message frame which is often not a negligible overhead.

The communication hardware model we use is slightly different from that for a general-purpose shared-memory parallel computer. The applications are not predictable in a general-purpose computer, and the division of address space separately for local and interprocessor communication cannot be done in advance when the computer is designed. In embedded systems, not all of the memory is shared, since shared memory only stores data communicated between PEs. Unlike a shared-memory parallel computer, embedded systems seldom achieve parallelism with a granularity finer than the process level. For instance, suppose there is a process which computes a fast Fourier transform (FFT). If no single CPU can implement the FFT in software to meet the performance requirements, we would use an ASIC to do the FFT rather than execute the FFT in parallel on several CPUs, because an ASIC may have higher performance and lower cost than the mechanism for fine-grain parallel processing. A general-purpose computer should be flexible and is unable to use an ASIC for all special functions, even though an ASIC may have higher performance and lower cost than the mechanism for fine-grain parallel processing. At the process level the interprocessor communication occurs only for the input and output data, so most of process computation will not be affected by bus activities. It is not necessary to handle problems like cache coherence in embedded systems.

The terms such as *shared memory* and *message passing* in parallel computer design terminology can be applied to either the *progam abstraction* level or the *hardware implementation* level. A message passing hardware can implement a shared memory program model, and vice versa. Using these terminologies in our system specification, the task graphs use message passing model in the progam abstraction level, and the shared memory communication is adopted in the hardware implementation level.

3.3.2 Communication Characterization

There are various types of PEs with different bus interfacing techniques. We characterize the bus interface schemes as follows. For each message, we create a **sending process** right after process P_1 to send the output data of P_1 to the shared memory, and a **receiving process** right before process P_2 to receive the data from the shared memory. The sending process is allocated to the same PE as P_1, and the receiving process is on the same PE as P_2. A **communication process** is either a sending process or a receiving process. We call a normal process in the original task specification an **application process**, whose existence is independent of the system architecture. Unlike an application process, a communication process needs to be allocated on not only a PE but also on a bus. The following properties should be identified for a communication process or a communication link:

- The **communication time** of a communication process is the time spent on finishing an uninterrupted data transfer. It is proportional to the size of the message, the PE speed, and the bus speed which depends on the width (8-bit, 32-bit, etc.) of the bus, the speed of the shared memory. The bus arbitration overhead should be incorporated into the communication time. The communication time on a bus is analogous to the computation time of processes on a PE.

- Whether the communication will interfere with the computation depends on the type of PE. The AT&T DSP32C DSP processor, for example, has a on-chip direct memory access (DMA) controller which can transfer data without processor intervention if the on-chip memory is large enough for computation, including instructions and data. The Motorola DSP96002 processor has dual ports, one of which can be used for independent communication, even though on-chip memory is not enough. Note that a DMA controller without on-chip memory can not really separate computation and communication, because a DMA transfer holds the local bus and prevents the CPU from fetching instructions.

- If there is a dual-port buffer between a PE and a bus, the sending or receiving process on the PE can be spared. Another PE can directly drop its data on the buffer instead of shared memory, so the PE with a buffer can read the data without using the bus. For example, the Motorola 6845 video controller has internal registers which other CPUs can directly read or write without interrupting the controller's operation. A dual-port memory can be added as a buffer for a CPU.

- Whenever a PE is connected to a bus, an associated cost is added to the total system cost. The cost includes bus interface logic and extra bus length. We assume the worst-case for such a cost has been given for each PE type and each bus type.

3.3.3 Message Allocation and Bus Scheduling

When a data dependency crosses PE boundary, we define the block of data for transfer as a **message**. In addition to allocation of processes onto PEs, the allocation of communication is a mapping of messages onto buses.

In bus communication, at most one master (PE) can utilize a bus at a time. When more than one PE want to send or receive a message through a bus, it is necessary to schedule the communication on the bus. A bus always transfers the message with the highest priority on the bus. A simple implementation scheme may assign a fixed priority for each PE on a bus. If a PE has higher priority than the others, all the sending and receiving processes will have higher priority than the communication processes on the other PEs. A more complex scheme, used in FutureBus and NuBus [5], lets a PE select its own priority during bus arbitration. Priorities can be assigned to each individual message rather than a whole PE by this method, which gives more scheduling flexibility. However, the second scheme requires more bus interface logic and more bus arbitration delay. We believe the first scheme is more common in practical design due to its simplicity.

A communication process actually uses two resources: a bus and a PE. In addition to scheduling the communication on the bus, we also need to schedule the communication processes on the PE, especially when the PE cannot perform communication and computation in parallel. In this case, the priorities are ordered for all the processes, including both application processes and communication processes. However, we assume that a communication process is non-preemptive on a CPU by considering the following situations:

- Preemption is often implemented by interrupts. Most existing microprocessors sample the interrupt signal only at the end of a bus cycle. When a communication process is unable to access a busy bus, the bus cycle is extended until it is granted the bus. The CPU cannot be interrupted in the middle of a bus cycle.

- If the communication is handled by a DMA controller, the CPU can use software to decide when to initiate the DMA transfer, so the initiation of the communication is priority schedulable. Once the DMA controller starts a block transfer, it will hold the local bus as if it has the highest priority on the local bus. Unless the DMA controller releases the local bus, the CPU is unable to fetch instructions and execute any other process.

- Preemption causes some performance overhead which is often acceptable and can be easily merged into the computation time on a single CPU. Nevertheless, a communication process consumes both CPU time and bus bandwidth. When a communication process is preempted, the CPU it is allocated to will swap the process, at the same time the bus will swap the granted PE, which may in turn cause the swap of processes on another CPU. In this case, the preemption overhead increases because it occurs on more than one PE as well as the bus, and the performance analysis problem becomes very complicated due to the interaction of several resources.

We model communication processes as non-preemptive processes. In other words, the initiation of a communication process follows its priority assignment, but once it is initiated, it will continue until finished on the CPU, even though some other processes with higher priority come later.

4

PERFORMANCE ANALYSIS

4.1 OVERVIEW

This chapter describes a new, efficient analysis algorithm [111, 112] to derive
tight bounds on the worst-case response time required for an application task
executing on a heterogeneous distributed system. Tight performance bounds
are essential to many co-synthesis algorithms [106, 26, 30]. Co-synthesis requires performance estimation techniques to select suitable hardware components and determine how different allocation and scheduling of processes can
affect the system performance. The performance estimation algorithm needs
to be efficient in order for the co-synthesis algorithm to quickly explore various architectures in a large design space. Given the computation time of the
uninterrupted execution of processes, the allocation of processes, and the priority assignment for process scheduling, the goal of our analysis algorithm is to
statically estimate the worst-case delay of a task.

These bounding algorithms are valid for the general problem model presented
in Chapter 3. Compared with the previous work discussed in Section 2.2, the
algorithms have the following advantages:

- The problem specification can contain multiple tasks with independent
  periods.

- It considers the data dependencies between processes.

- It uses upper and lower bounds instead of constants on both task periods
  and process computation times.

- It can handle preemption and task pipelining.

- It utilizes analytic approach which is more efficient than enumeration or simulation.

Such a model requires more sophisticated algorithms, but leads to more realistic results than previous work.

Unlike circuit area or component prices which can be approximated by monotonic additive cost attributes [33], timing behavior is extremely complex and requires special analysis methods. While scheduling and allocation of a single nonperiodic task is NP-complete, its performance analysis is polynomially solvable, given an allocation and schedule of processes. Therefore, performance analysis of a single nonperiodic task is seldom treated separately in previous work. On the other hand, for the periodic concurrent tasks running on a distributed system, we can show that even the analysis problem is NP-hard, based on a result proved by Leung and Whitehead [56].

Theorem 4.1: Given the task graphs (multiple periodic tasks and data dependencies between processes) and an embedded system architecture (the hardware engine, schedule and allocation of processes), the problem to decide whether the deadline of each task is satisfied is NP-hard.

Proof:

Leung and Whitehead [56] proved that deciding whether a priority assignment schedule is valid for an asynchronous system on a single processor is an NP-hard problem. An asynchronous system is defined to be a set of processes where each process P_i has a period p_i, a deadline d_i, a computation time c_i, and an initial invocation time s_i.

We prove that our analysis problem is NP-hard by showing that the asynchronous system analysis problem is polynomial-time reducible [72, 24] to our analysis problem. For each process P_i in the asynchronous system, create a task graph τ_i as follows:

- Task τ_i contains P_i, another process Q_i, and an directed edge from Q_i to P_i.

- The computation time of Q_i is s_i, and the computation time of P_i is c_i.

- The period of τ_i is p_i and the deadline of τ_i is $d_i + s_i$.

Create an embedded system architecture as follows:

- Allocate each Q_i on a different PE such that only Q_i is executed on that PE.

- Allocate all the processes P_i's in the original asynchronous system on the same CPU.

- Use the priority assignment for the original asynchronous system to schedule the processes P_i's.

Under this embedded system architecture, the deadline of task τ_i is satisfied if and only if the deadlines of process P_i in the original asynchronous system is satisfied. Because the analysis problem of an asynchronous system is NP-hard, the performance analysis of task graphs on an embedded system architecture is NP-hard.   ■

We adopt analytical methods which provide *conservative* delay estimates—they are strict upper bounds on delay. Most algorithms for hard real-time distributed systems enumerate all the occurrence of processes in a cycle with a length equal to the least common multiple of all the task periods to avoid the difficulties of periodic behavior. This approach is not practical for co-synthesis, as mentioned in Section 2.2. We need new delay estimation method for co-synthesis.

In Section 4.2 we use a *fixed-point iteration* technique to calculate the exact worst-case response time of processes without data dependencies. Then in Section 4.3 we propose *phase adjustment* technique, combined with a longest-path algorithm, to take data dependencies into consideration. We present *separation analysis* in Section 4.4 to detect whether two processes may overlap their execution and ignore the interaction between disjoint processes. Section 4.5 describes our overall performance analysis algorithm. Before Section 4.6 we assume that the period of a process is the same as that of the task it belongs to; this is not true in general, and we make the necessary modifications for *period shifting* effects—the actual invocation period of a process may change due to data dependencies. We discuss in Section 4.7 how our algorithm can still be valid under *task pipelining*—the delay of a task is longer than its period.

To ease the discussion, communication delay is ignored in this chapter. Chapter 6 will extend the algorithms in this chapter to handle communication analysis.

4.2 FIXED-POINT ITERATION

In this section, we elaborate on an existing method for computing the worst-case response time of independent processes. Suppose $P_1, P_2, \ldots$ are a set of priority-ordered processes allocated to the same CPU, with P_1 being the process with the highest priority. There is no data dependence between the processes so their invocation are independent. According to the task graph model in Section 3.1, it is equivalent to the case in which each task contains exactly one process whose period is the same as the task period. For a process P_i, its minimum period is $p_i = p_i^{lower} > 0$, and its longest computation time on the CPU is $c_i = c_i^{upper} > 0$. Let the worst-case response time from an invocation of P_i to its completion be w_i, then w_i is the smallest positive root of Equation (2.2), which is repeated below from Section 2.1.2:

$$x = g(x) = c_i + \sum_{j=1}^{i-1} c_j \cdot \lceil x/p_j \rceil$$

Although the iteration technique to solve this nonlinear equation has been mentioned by Sha et al. [88], they did not analyze the convergence of this method, and the initial value for the iteration is not tight enough. We restate the method as a *fixed-point iteration* [15] technique:

1   if $(\sum_{j=1}^{i-1} c_j/p_j \geq 1)$

2   return FAIL;

3    $x = \lceil c_i/(1 - \sum_{j=1}^{i-1} c_j/p_j) \rceil$;

4   while $(x < g(x))$

5        $x = g(x)$;

6   return x;

As mentioned in Section 3.1.1, without loss of generality, all numbers are assumed to be integers. It is assumed that $(1 - \sum_{j=1}^{i-1} c_j/p_j) > 0$; otherwise, the schedule must be infeasible and such an allocation should be given up by the synthesis algorithm because the processor utilization $U \equiv \sum_{j=1}^{i} c_j/p_j > 1$ and no feasible schedule can exist. Fixed-point iteration cannot be applied to arbitrary equations because the convergence may not be guaranteed. Nonetheless, we can prove that the value of x must converge to w_i in a finite number of steps.

Theorem 4.2: The fixed-point iterations for Equation (2.2) will converge to w_i. The number of iterations will not exceed the number of the multiples of the periods within the interval

$$[\frac{c_i}{1 - \sum_{j=1}^{i-1} c_j/p_j}, \frac{\sum_{j=1}^{i} c_j}{1 - \sum_{j=1}^{i-1} c_j/p_j}]$$

Proof:

$$c_i + \sum_{j=1}^{i-1} c_j \cdot (x/p_j) \leq g(x) = c_i + \sum_{j=1}^{i-1} c_j \cdot \lceil x/p_j \rceil \leq c_i + \sum_{j=1}^{i-1} c_j \cdot (x/p_j + 1)$$

$$c_i + \sum_{j=1}^{i-1} c_j \cdot (w_i/p_j) \leq w_i = g(w_i) \leq c_i + \sum_{j=1}^{i-1} c_j \cdot (w_i/p_j + 1)$$

$$\frac{c_i}{1 - \sum_{j=1}^{i-1} c_j/p_j} \leq w_i \leq \frac{\sum_{j=1}^{i} c_j}{1 - \sum_{j=1}^{i-1} c_j/p_j}$$

Initially, $x = \lceil c_i/(1 - \sum_{j=1}^{i-1} c_j/p_j) \rceil \leq w_i$. Apparently, $g(x)$ is an increasing function, so if $x < w_i$, $g(x) \leq g(w_i) = w_i$. When $x = w_i$, the iterations will stop because $x = g(x)$. As a result, it is impossible for the value of x to exceed w_i. Function $g(x)$ can increase only at a multiple of a period and keeps constant between two multiples. To continue an iteration, x must increase over a multiple of a period, otherwise $g(x)$ will not increase and $x = g(x)$ which terminates the iterations. Since there are only a finite number of multiples of the periods between $c_i/1 - \sum_{j=1}^{i-1} c_j/p_j < w_i$ and $(\sum_{j=1}^{i} c_j)/(1 - \sum_{j=1}^{i-1} c_j/p_j) > w_i$, x must converge to w_i in a finite number of steps. ∎

Example 4.1: Suppose $p_1 = 5, c_1 = 1, p_2 = 37, c_2 = 3, p_3 = 51, c_3 = 16, p_4 = 134, c_4 = 42$. By fixed-point iteration, we only need 4 steps to know $w_4 = 128$. The x-values during iterations are 104, 120, 126 and 128.

Note it needs 31 steps by enumerating the multiples of the periods [53] to get the solution. During fixed-point iterations, the initial value 104 has skipped 23 multiples. When x is increased from 104 to 120, for example, 4 multiples are jumped over. The iteration method is more efficient than trying all multiples of the periods.

It needs 8 steps to converge by using the iteration scheme mentioned by Sha et al. [88], because the computation time $c_4 = 42$ is used as the initial x value which is less tight.

In the restricting case where the deadline is equal to the period, by rate monotonic analysis in formula (2.1) the processor utilization

$$U = \frac{1}{5} + \frac{3}{37} + \frac{16}{51} + \frac{42}{134} = 0.908 > 4(2^{1/4} - 1) = 0.76$$

and the schedule may be misjudged to be infeasible, while by fixed-point iteration the deadlines are satisfied. We can see that rate monotonic analysis is very pessimistic. In addition to this, formula (2.1) becomes invalid when the deadline is smaller than the period.   ■

4.3   PHASE ADJUSTMENT

We will next develop a new algorithm which approximates the worst-case delay through a task graph executing on multiple PEs. This algorithm takes into account complications caused by data dependencies in a task graphs. We propose *phase adjustments* for fixed-point iteration in a longest-path algorithm. If a preempting process P_j has occurred some time earlier than the current process P_i in a path, P_j may not be able to occur immediately and its occurrence is constrainted by a *phase* which is the soonest time for its next request relative to the request time of the preempted process P_i.

Example 4.2: The example of Figure 4.1 illustrates the effects of data dependencies on delay. If we ignore data dependencies between P_2 and P_3, as is done by solving Equation (2.2), their worst-case response times are 35 and 45, respectively. But the worst-case total delay along the path from P_2 to P_3 is

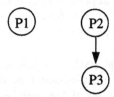

two task graphs

process	period	computation time
P_1	80	15
P_2	50	20
P_3	50	10

process characteristics

Figure 4.1 An example of worst-case task delay estimation. The three processes share the same CPU and P_1 has highest priority.

actually 45 instead of 80, the sum of individual process delays, because of the following two conditions:

1. P_1 can only preempt either P_2 or P_3, but not both in a single execution of the task.

2. It is impossible for P_2 to preempt P_3.

■

If the delay of a single path is the sum of the delays in each node along the path, we can use an $O(V + E)$ longest-path algorithm [105, 24] to calculate the longest delay of a task graph—the longest delay is the length of the longest path from the START node to the END node. In a single-source longest-path algorithm, each node has an attribute called *longest path estimate*, which is a lower bound of the longest path length from a given source to the node. In a task graph, the source is the START node. If the length of each path is nonnegative, as in the case for delay calculation, initially all longest path estimates are set to zero. When an edge (u, v) is *relaxed*, the longest path estimate of node v is updated. For an acyclic graph, if the edges are updated according to a topologically sorted order, the final values of the longest path estimates are the length of the longest paths.

In our case, the delays through several paths are not independent. To estimate task delay more accurately, we must consider combinations of process activations to avoid impossible preemption in the specified dataflow.

4.3.1   A Modified Longest Path Algorithm

We use *phase adjustments* in a longest-path algorithm to deal with the first condition mentioned in Example 4.2. If process P_j has higher priority than P_i on the same CPU, and is assumed to occur some time before the request time of process P_i, the value of *phase* ϕ_{ij}^r is the earliest time for the next request of process P_j relative to the request time of process P_i. We have developed the algorithm LatestTimes, listed in Figure 4.2, to compute the **latest request time** and **latest finish time** of each process P_i in a task. These times are relative to the start of the task, so the latest finish time of the END node of the task is the worst-case delay of the whole task.

LatestTimes is a modified longest-path algorithm. Similar to a longest path estimate, each process P has a **latest request time estimate** $P.request$ and a **latest finish time estimate** $P.finish$; their initial values are set to zero at line 6, and the final values of $P.request$ and $P.finish$ are the latest request time and latest finish time. The value of $P.finish$ is $P.request$ plus the worst-case response time of P, as described at line 11. The value of $P_i.request$ is updated at line 16 when a dependency edge from P_k to P_i is relaxed in lines 14–17.

In task delay analysis, a path delay is the sum of the node weights, which are the worst-case response times of processes. In normal longest path problems, the weight on a node or an edge is constant and given in advance, but in LatestTimes the worst-case response time of a process is computed dynamically when the process is visited. Line 10 of Figure 4.2 calculates the worst-case response time of a process P_i as follows:

- **Non-preemptive process:** If the process is allocated on hardware, or it is a dummy process, the response time w_i is the computation time of P_i, because the execution of such a process will not be interrupted by other processes.

1 LatestTimes(a task graph G)

2 /* Compute $latest[P_i.request]$ and $latest[P_j.finish]$ for all P_j

3   in a task graph G. */

4 {

5   for (each process P_i) {

6       $latest[P_i.request] = 0$;

7   for (each process P_j) $\phi_{ij}^r = 0$;

8   }

9   for (each process P_i in topologically sorted order) {

10   w_i = the worst-case response time of P_i with phase adjustment by ϕ_{ij}^r;

11       $latest[P_i.finish] = latest[P_i.request] + w_i$;

12   Calculate the phases ϕ_{ij}^f relative to $latest[P_i.finish]$ for each j;

13   for (each immediate successor P_k of P_i) {

14          $\delta = latest[P_k.request] - latest[P_i.finish]$;

15   if $(latest[P_k.request] < latest[P_i.finish])$

16             $latest[P_k.request] = latest[P_i.finish]$;

17   Update the phases ϕ_{kj}^r for each j according to ϕ_{ij}^f and δ;

18   }

19   }

20}

Figure 4.2 The algorithm LatestTimes for finding the latest request time and the latest finish time from a starting node in a task graph.

- **Preemptive process:** If the process is allocated on a CPU, we need to apply fixed-point iteration for w_i by (2.2). However, the terms $\lceil x/p_j \rceil$ in $g(x)$ shown in Equation (2.2) are modified into

$$\lceil (x - \phi_{ij}^r)/p_j \rceil \tag{4.1}$$

where ϕ_{ij}^r is the phase of P_j relative to the request time of P_i. In other words, the variable x is adjusted by phase ϕ_{ij}^r.

4.3.2   Phase Updating

In addition to updating the latest request time estimates and latest finish time estimates, LatestTimes also updates the phases ϕ_{ij}^r and ϕ_{ij}^f for each pair of processes P_i and P_j. Initially, the ϕ_{ij}^r's are set to zero for all i, j at line 7 of Figure 4.2. Then they are updated at line 17 when their immediate predecessors are visited. After w_i is computed, line 12 calculates ϕ_{ij}^f's, the phases relative to $latest[P_i.finish]$ by the following rules:

- If P_j preempts P_i,

$$\phi_{ij}^f = (\phi_{ij}^r - w_i) \bmod p_j \tag{4.2}$$

We subtract w_i from ϕ_{ij} because $latest[P_i.finish] = latest[P_i.request] + w_i$.

- If P_j does not preempt P_i, either because it is allocated to a different PE or because its priority is lower than P_i,

$$\phi_{ij}^f = \max(\phi_{ij}^r - w_i, 0) \tag{4.3}$$

If $\phi_{ij}^f < 0$, we set it to zero for the worst-case timing.

Note that ϕ_{ij} is always in the range $[0, p_j)$.

In Figure 4.2, line 17 uses ϕ_{ij}^f and δ calculated at line 14 to update ϕ_{kj}^r for each immediate successor P_k of P_i as follows:

- If P_i is the first visited immediate predecessor of P_k, $\phi_{kj}^r = \phi_{ij}^f$ for each j, because $latest[P_k.request] = latest[P_i.finish]$ in this case.

- If $\delta \geq 0$, there is a slack between $latest[P_i.finish]$ and $latest[P_k.request]$. From the highest priority to the lowest, if $\phi_{ij}^f < \phi_{kj}^r$, increase ϕ_{ij} by δ and reduce δ by c_j until δ becomes negative. Using the slacks to reduce the phase value obtains tighter estimation for w_i's, according to the adjustment formula (4.1) and Equation (2.2). Then

$$\phi_{kj}^r = \min(\phi_{kj}^r, \phi_{ij}^f)$$

We choose the smaller phase which may give longer delay to P_k for the worst case.

- If $\delta < 0$, increase ϕ_{kj}^r by $|\delta|$ similarly and choose the smaller phase.

The updated ϕ_{kj}^r values are used to adjust the phases in fixed-point iteration to get a more accurate value for w_k when P_k is visited.

Processes P_i's or P_k's visited by LatestTimes belong to the same task graph. However, the worst-case delay of the task graph may be affected by processes P_j's from the other tasks, if those processes share a CPU. For instance, the delay of the task graph composed of P_2 and P_3 in Figure 4.1 can be lengthened by P_1 in another task.

Example 4.3: For the example in Figure 4.1, $\phi_{21}^r = 0$ initially. When the algorithm LatestTimes visits P_2, by solving the equation

$$x = g(x) = 20 + 15 \cdot \lceil x/80 \rceil$$

we get $latest[P_2.finish] = latest[P_3.request] = w_2 = 35$ and

$$\phi_{31}^r = \phi_{21}^f = (\phi_{21}^r - w_2) \bmod p_1 = -35 \bmod 80 = 45$$

If we know P_2 will not preempt P_3 because it is P_3's predecessor in the task graph, and solve the equation with phase adjustment

$$x = g(x) = 10 + 15 \cdot \lceil (x - 45)/80 \rceil$$

we get $w_3 = 10$ and $latest[P_3.finish] = 35 + 10 = 45$, which is the worst-cast task delay we expect. ∎

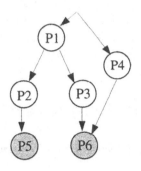

a task graph

case	P_1	P_2	P_3	P_4	P_5	P_6
1	[12, 15]	[5, 6]	[20, 25]	[8, 10]	[10, 10]	[10, 10]
2	[12, 15]	[20, 25]	[5, 6]	[8, 10]	[10, 10]	[10, 10]
3	[12, 15]	[20, 25]	[5, 12]	[8, 10]	[10, 10]	[10, 10]
4	[12, 15]	[20, 25]	[5, 6]	[30, 35]	[10, 10]	[10, 10]

process computation times

Figure 4.3 Each computation time entry $[l, u]$ represents a lower bound l and an upper bound u. Four different cases are listed.

4.4 SEPARATION ANALYSIS

In order to handle the second condition mentioned in Example 4.2, we need to know whether the execution of two processes will overlap or not. If two processes do not have overlapping execution windows, they cannot preempt each other. While it is easy to know that a process will not preempt its predecessors or successors like the case for P_2 and P_3 in Figure 4.1, it is not obvious how to decide how the execution of two processes overlap in disjoint paths of a task graph.

Example 4.4: Figure 4.3 gives four different combinations of process computation times. In this example, by allocating P_1, P_2, P_3, P_4 and P_5 to different PEs, we consider only the variation in computation time and ignore the variation due to preemptions. P_5 and P_6 are on the same CPU where P_5 has higher priority.

In case 1 and case 2, P_5 will not preempt P_6 because their execution window is separated. But in case 3 and case 4, it is possible for P_5 to preempt P_6 and we need to take this into consideration for the worst-case task delay estimation. ∎

4.4.1 Max Constraint Model

We want to know the separation between two processes in the same task. To calculate the separation value between two processes in the same task according to the response times of all the processes, we need to be able to handle two phenomena. First, the response time of a process is not a fixed constant, and can be represented by a lower bound and an upper bound. There is a variation in both the computation time and the period of a process and the number of preemptions can also vary from period to period. Second, the relationship between the start of a process and its predecessors in the task graph should be modeled by **max constraints**—the initiation time of a process is computed by a max function of the finish times of its immediate predecessors in the task graph.

According to our problem model in Chapter 3, the dependency edges in a task graph behave like max constraints, as described in Section 2.6. McMillan and Dill's $O(ne)$ algorithm [66], listed in Figure 4.4, can deal with max constraints and calculate pairwise maximum event separations. The maximum separation from event a to event b is the largest time difference between these two events for all possible sets of delay values which satisfy both lower bounds and upper bounds on the constraints. If the maximum separation from event a to event b is negative, it means that event b must occur before event a no matter what the actual delay values are. However, in their algorithm the delay along a path is the sum of the bounds of individual constraints, which in not true in our case, as mentioned in Example 4.2. We need a method which utilizes LatestTimes to derive delay bounds between processes in a task graph, where a delay along a path is not the sum of individual process delays.

4.4.2   Delay Bounds between Processes

For each process P, we consider two time instants—$P.request$ is the request time of P relative to the beginning of the task graph, and $P.finish$ is the time at which P finishes its execution. We want to derive the delay bounds between a pair of time instants in a path of a task graph. Given two time instants x and y, $upper[x,y]$ is an upper bound on the delay from x to y and $lower[x,y]$ is a lower bound on the delay from x to y.

The upper bounds can be calculated as follows. Let G_i be a subgraph composed of P_i and all its successors. After calling LatestTimes(G_i) we can obtain the upper bounds from a request time by assigning

$$upper[P_i.request, P_j.request] = latest[P_j.request]$$

$$upper[P_i.request, P_j.finish] = latest[P_j.finish] \qquad (4.4)$$

for all successors P_j of P_i. The upper bound from a finish time can be derived according to the upper bounds from the request time, that is,

$$upper[P_i.finish, P_j.request] = \max_k upper[P_k.request, P_j.request]$$

$$upper[P_i.finish, P_j.finish] = \max_k upper[P_k.request, P_j.finish]$$

where P_k is a immediate successor of P_i.

The lower bounds can be obtained in a similar way. We can modify the algorithm LatestTimes in Figure 4.2 as follows to obtain the earliest request time $earliest[P_i.request]$ and the earliest finish time $earliest[P_i.finish]$:

- Replace $latest[\cdot]$ with $earliest[\cdot]$.

- Replace the ceiling operators $\lceil \cdot \rceil$ with the floor operators $\lfloor \cdot \rfloor$ in Equation (2.2) for fixed-point iteration.

- Let $c_i = c_i^{lower}$ and $p_i = p_i^{upper}$ for the best-case delay estimation.

- At line 12 of Figure 4.2, set ϕ_{ij}^f to 0 if P_j are not allocated to the same CPU as P_i. Otherwise, calculate ϕ_{ij}^f in a way similar to that in Section 4.3 but keep it in the range $(-p_j, 0]$. In other words, subtract p_j from it if it is greater than 0 and set it to 0 if it is below $-p_j$.

- At line 17, make ϕ_{kj}^r equal to ϕ_{ij}^f only when $earliest[P_k.request] = earliest[P_i.finish]$. Otherwise, leave ϕ_{kj}^r unchanged.

Suppose the modified algorithm is EarliestTimes. We can call EarliestTimes(G_i) and let

$$lower[P_i.request, P_j.request] = earliest[P_j.request]$$

$$lower[P_i.request, P_j.finish] = earliest[P_j.finish]$$

The lower bound from a finish time can be derived according to the lower bounds from the request time, that is,

$$lower[P_i.finish, P_j.request] = \max_k lower[P_k.request, P_j.request]$$

$$lower[P_i.finish, P_j.finish] = \max_k lower[P_k.request, P_j.finish]$$

where P_k is an immediate successor of P_i.

4.4.3 Maximum Separation Algorithm

The algorithm MaxSeparations, listed in Figure 4.5, is a modified version of McMillan and Dill's maximum separation algorithm. It utilizes the delay bounds calculated by LatestTimes and EarliestTimes to derive the maximum separations between two time instants. A bound between two time instants is not equivalent to the maximum separation for two reasons:

- LatestTimes and EarliestTimes only gives a bound between processes in the same path of a task graph. The maximum separation between two processes in different paths can only be derived by MaxSeparations.

```
1  P(i, j)
2  {
3     if (p_ij has not already been defined) {
4        if (i = j)
5           p_ii = 0;
6        else
7           p_ij = min_{k∈succs(j)}(P(i, k) − l_{j,k});
8     }
9     return p_ij;
10 }
```

```
11 S(i, j)
12 {
13    if (s_ij has not already been defined) {
14       if (i = j)
15          s_ii = 0;
16       else
17          s_ij = min(max_{k∈preds(i)}(S(i, k) + u_{kj}), P(i, j));
18    }
19    return s_ij;
20 }
```

Figure 4.4   McMillan and Dill's maximum separation algorithm. l_{ij} and u_{ij} are the lower bound and upper bound on the delay between i and j. The function $S(i, j)$ returns the maximum separation from i to j.

```
1  MaxSeparations(a process Pᵢ)
```

1 MaxSeparations(a process P_i)

2 /* Compute $maxsep[P_i.request, P_j.finish]$ for all P_j in the same task graph */

3 {

4   for (each process P_j in topologically sorted order) {

5   Enqueue(Q, P_j);

6        $tmp = \infty$;

7   while (Q is not empty) {

8   P_k = Dequeue(Q);

9   for (each immediate predecessor P_l of P_k) {

10   if (P_l is a predecessor of P_i)

11             $tmp = \max(tmp, upper[P_k.request, P_j.finish]$

12             $-lower[P_l.finish, P_i.request])$;

13   if (P_j is a predecessor of P_i)

14             $tmp = \min(tmp, -lower[P_j.finish, P_i.request])$;

15   if ($maxsep[P_i.request, P_l.finish] == -lower[P_l.finish, P_i.request]$)

16   continue;

17   if ($P_l \notin Q$) Enqueue(Q, P_l);

18   }

19   if ($tmp == -lower[P_j.finish, P_i.request]$) break;

20   }

21       $maxsep[P_i.request, P_j.finish] = tmp$;

22   }

23 }

Figure 4.5 The algorithm MaxSeparations for finding the maximum separations from P_i to all the processes in the same task graph.

- The bound between two processes in the same path may not be the maximum separation between them. The maximum separation is determined by composite effects from different paths.

McMillan and Dill's maximum separation algorithm [66], listed in Figure 4.4, is a recursive depth-first search procedure that calls a second procedure for finding shortest paths. The function $P(i, j)$ gives a possible separation bound induced by the lower bounds in the paths from event i to event j. The maximum separation from i to j is also bounded by the sum of the maximum separation from i to k and the upper bound from k to j, where k is a predecessor of j in the event graph. Line 17 of Figure 4.4 is the major step in the algorithm. It can be stated as follows in our notation:

$$maxsep[i, j] = \min(\max_k(maxsep[i, k] + upper[k, j]), -lower[j, i]) \qquad (4.5)$$

However, two phenomena require us to use different approaches to implement a similar idea:

- The delay intervals are associated with the nodes (processes) in a task graph, while the delay intervals are associated with the edges (constraints) in an event graph.

- The total delay along a path is not the sum of node delays.

Given a source node i in McMillan and Dill's algorithm, the maximum separation $maxsep[i, j]$ can be calculated only after $maxsep[i, k]$ is known for all predecessors k of j through recursive calls. MaxSeparations uses a non-recursive version of implementation; in Figure 4.5, line 4 searches the task graph in topologically sorted order for the same reason. Similar to formula (4.5), we apply the $\max_k$ and min operators at line 11 and 14, respectively. However, when a path delay is not the sum of bounds, it is not accurate enough to consider only the immediate predecessors and use the formula $maxsep[i, k] + upper[k, j]$ to calculate upper bounds. Instead, we try to consider all the predecessors of P_j by a backward breadth-first-search in lines 7–20. The breadth-first-search is trimmed at line 15 and terminated at line 19 when the lower bounds determine the maximum separation value and it is not necessary to trace backwards further to know the upper bounds along a path.

4.5   ITERATIVE TIGHTENING

Algorithm MaxSeparations gives the maximum separation from a request time to a finish time. The maximum separation between two request times can be calculated as described below:

$$maxsep[P_i.request, P_j.request] = \max_k maxsep[P_i.request, P_k.finish]$$

where P_k is an immediate predecessor of P_j. The maximum separations are used to improve delay estimation in LatestTimes or EarliestTimes as follows:

■   In LatestTimes, if

$$maxsep[P_i.request, P_j.finish] \leq 0 \text{ or}$$

$$maxsep[P_j.request, P_i.finish] \leq 0, \tag{4.6}$$

the execution of P_i and P_j will not overlap and the corresponding terms are eliminated from the function $g(x)$ in Equation (2.2) when the worst-case response time w_i is computed at line 10 of Figure 4.2. As an example, suppose there are three processes P_1, P_2, and P_3 with higher priority than P_4 on the same CPU. According to Equation (2.2), w_4 is solved from the following equation

$$x = g(x) = c_4 + \sum_{j=1}^{3} c_j \cdot \lceil x/p_j \rceil$$

In case we get $maxsep[P_2.request, P_4.finish] \leq 0$, the equation can be reduced to

$$x = g(x) = c_4 + c_1 \cdot \lceil x/p_1 \rceil + c_3 \cdot \lceil x/p_3 \rceil$$

and a tighter value for w_4 can be derived.

■   In LatestTimes, if $maxsep[P_j.request, P_i.request] < 0$,

$$\phi_{ij}^r = \max(\phi_{ij}^r, -maxsep[P_j.request, P_i.request]) \tag{4.7}$$

for phase adjustment.

- In EarliestTimes, if $maxsep[P_j.request, P_i.request] <= 0$,

$$\phi_{ij}^{\tau} = \min(\phi_{ij}^{\tau}, maxsep[P_i.request, P_j.request] - p_j)$$

for phase adjustment.

We use maximum separations calculated by MaxSeparations to improve delay estimation in both LatestTimes and EarliestTimes. We need to call both LatestTimes to derive maximum separations in MaxSeparations. Therefore, we get successively tighter delay bounds and maximum separations iteratively. Initially, to be more pessimistic, we let the maximum separations be ∞. The iterative procedure, DelayEstimate, is described in Figure 4.6. The iterative delay estimation stops when the maximum separations do not change any more. We can prove that it must terminate.

Theorem 4.3: The delay estimation algorithm DelayEstimate must terminate in a finite number of steps.
Proof:
We shall first show by induction that during the iterative tightening:

1. the maximum separation values $maxsep[\cdot, \cdot]$ derived by MaxSeparations must be non-increasing;

2. the upper bound values $upper[\cdot, \cdot]$ calculated by LatestTimes must be non-increasing;

3. the lower bound values $lower[\cdot, \cdot]$ calculated by EarliestTimes must be non-decreasing.

Because the initial values of $maxsep[\cdot, \cdot]$ are ∞, in the first iteration, the new $maxsep[\cdot, \cdot]$ values cannot increase. Suppose in the kth iteration, the values of $maxsep[\cdot, \cdot]$ are non-increasing. In the $k + 1^{th}$ iteration, the values of $latest[\cdot]$ in Figure 4.2 must be non-increasing for the following reasons:

- If inequalities (4.6) hold in the kth iteration such that some terms can be eliminated from Equation (2.2), it is so in the $k + 1^{th}$ iteration too. The number of terms in Equation (2.2) will not increase for a certain w_i at line 10 of Figure 4.2. The value of w_i will not increase because of the change in the number of terms in Equation (2.2).

- According to formula (4.7), the phase values of ϕ_{ij}^r's will not decrease because of the change in $maxsep[\cdot, \cdot]$. If the phase values do not decrease, by the phase adjustment rule in formula (4.1), the values of w_i's will not increase.

- The non-decreasing property of w_i's will not increase the phase adjustment values, according to formulas (4.2) and (4.2). This avoids the increase of w_i's in turn.

- If the values of w_i's cannot increase, the values of $latest[\cdot]$ do not increase either, as seen in line 11 of Figure 4.2.

The upper bounds $lower[\cdot, \cdot]$ are derived from the values of $latest[\cdot]$ in formula (4.4), so the upper bound values are non-increasing. Similar arguments can be applied to show that the lower bounds $lower[\cdot, \cdot]$ are non-decreasing.

In MaxSeparations, from line 11 and line 14 of Figure 4.5, the values of $maxsep[\cdot, \cdot]$ are non-increasing as long as $upper[\cdot, \cdot]$ are non-increasing and $lower[\cdot, \cdot]$ are non-decreasing.

Because both $upper[\cdot, \cdot]$ and $lower[\cdot, \cdot]$ are finite and nonnegative, $maxsep[\cdot, \cdot]$ cannot be $-\infty$. Since $maxsep[\cdot, \cdot]$ cannot decrease forever, at some point they must remain unchanged and algorithm DelayEstimate terminates. ∎

We may also set a limit on the number of iterations if faster delay estimation is desirable. Since the delay estimate is tightened at each step, the estimation process can be stopped at any time. However, in the experiments of Section 4.8 no limit was set, because in practice the number of iterations is very small. The final value of either $latest[END.finish]$ or $maxsep[START.request, END.finish]$ is the worst-case delay estimation for a task.

4.6   PERIOD SHIFTING

So far, we have assumed the period of a process is the same as that of the task to which the process belongs. While this is true for a process with no predecessors in a task graph, this assumption is not accurate in general. The delay for the predecessors may vary from period to period, making the request period of a process different from the period of the task.

```
1 DelayEstimate(a task graph G)
2 {
3    maxsep[·, ·] = ∞;
4    step = 0;
5    do {
6       for (each Pᵢ in G) {
7          EarliestTimes(Gᵢ);
8          LatestTimes(Gᵢ);
9       }
10      for (each Pᵢ)
11         MaxSeparations(Pᵢ);
12      step++;
13   } while (maxsep[·, ·] is changed and step < limit); }
```

Figure 4.6   The iterative procedures for tightening delay estimation.

Example 4.5: For the task graphs and a system implementation shown in Figure 4.7, if we consider the period of P_3 to be 70 which is the same as that of τ_2, the worst case delay of τ_3 should be 50. However, the worst case delay of τ_3 can be 80. Note a process cannot start before its request which may be activated by an external signal arrives, or the completion of its immediate predecessors in the task graph.   ■

Suppose there is no task pipelining so all the processes finish before their next requests. In Equation (2.2) the maximum number of requests for the processes P_j is $\lceil x/p_j \rceil$, where p_j is the period of the task containing P_j. Before any preemption from P_j occurs in a task graph, the term for the number of requests should be modified into

$$\lceil (x + latest[P_j.request] - earliest[P_j.request])/p_j \rceil$$

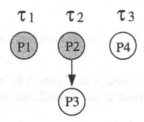

three task graphs

task	period	process	computation time
τ_1	150	P_1	30
τ_2	70	P_2	10
τ_3	110	P_3	30
		P_4	20

task characteristics

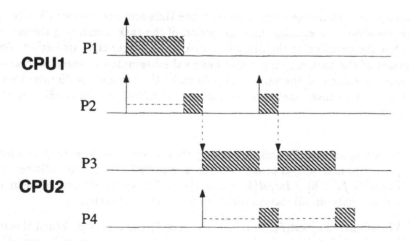

Figure 4.7 An example of the period shifting effect. An upward arrow represents the request of a process. A downward dotted arrow stands for a data dependency. P_1 has higher priority than P_2 on one CPU while P_3 has higher priority than P_4 on the other. The timing diagram shows that the worst case delay of τ_3 is 80.

We call such modification *period shifting*. Similarly, when we calculate the earliest times, the minimum number of requests should be

$$\lfloor (x - latest[P_j.request] + earliest[P_j.request])/p_j \rfloor$$

In the iterative tightening procedures in Figure 4.6, initially when no information about the earliest and latest times is available, we make

$$latest[P_j.request] - earliest[P_j.request] = p_j - c_j$$

for the worst case, because otherwise P_j may not finish before its next request. Later on, the values for $latest[P_j.request]$ and $earliest[P_j.request]$ are tightened step by step in DelayEstimate, and the period shifting is modeled more and more accurate.

4.7   TASK PIPELINING

Although we assume the worst-case response time and the computation time of a process should be smaller than the period of the task containing the process, we allow the deadline or the delay of a task to be longer than its period. Some processes of the task may not finish before the beginning of some processes in the next execution of the same task. To make the techniques discussed so far valid in spite of *task pipelining*, we require the following two conditions to be satisfied:

- If two processes P_i and P_j belong to the same task with a minimum period p and are allocated to the same CPU, it is not allowed that $latest[P_i.finish] > latest[P_j.request] + p$. The synthesis procedure should not generate an allocation which causes such a situation.

- We will avoid $latest[P_i.finish] > earliest[P_j.request] + p$. When this does happen, we implement a dummy process with a delay $latest[P_i.finish] - p$ between the start of the task and the initiation of P_j.

If the first requirement is not satisfied, P_i will delay the request time of P_j, which may in turn delay the request time of the next iteration of P_i further

Function	Description	lower bound	upper bound
checkdata	Park's example	32	1039
piksrt	Insertion Sort	146	4333
des	Data Encryption Standard	42302	604169
line	Line drawing	336	8485
circle	Circle drawing	502	16652
jpegfdct	JPEG forward discrete cosine transform	4583	16291
jpegidct	JPEG inverse discrete cosine transform	1541	20665
recon	MPEG2 decoder reconstruction	1824	9319
fullsearch	MPEG2 encoder frame search routine	43082	244305
matgen	Matrix generating routine	5507	13933

Table 4.1 The bounded computation time on i960 for several routines [57].

and there is a chance that the delay is getting longer and longer. If the second requirement is not satisfied, when $P_i = P_j$, the peak frequency of P_i may get too high due to period-shifting effect which endangers the deadlines of other tasks. As a matter of fact, such requirements are conservative. However, in most practical pipelined designs, different stages are allocated to different resources (PEs). The execution time of each stage is smaller than the period and it is unlikely that a process will overlap the next execution of the same stage, so these requirements are reasonable in practice.

4.8 EXPERIMENTAL RESULTS

The whole performance analysis algorithm was implemented in C++, and some experiments were performed on a Sun SS20 Sparc Workstation. Some examples in other literature are not suitable for comparison. For instance, the example by Ramamritham [80] did not use static allocation; the example by Peng and Shin [76] used synchronization to make three tasks equivalent to a single non-periodic task. Both examples have small periods and the LCM of the periods happen to be the largest period.

4.8.1 An Example Based on Li and Malik's Data

Function	Description	lower bound	upper bound
arccos	arc cosine	166	706
sqrt	square root	460	460
gran	random number generator	1128	1128
matmul	matrix multiplication	810	810
fft	fast Fourier transform	103688	103688

Table 4.2 The bounded computation time on DSP3210 for several routines [58].

CPU	priority-ordered processes
i960-1	piksrt, line, circle, jpegidct
i960-2	jpegfdct, matgen, fullsearch, checkdata
i960-3	des, recon
DSP3210	sqrt, arccos, matmul, gran, fft

Table 4.3 The allocation and schedule of the processes in Figure 4.8.

Li and Malik estimated the computation time for some real programs on i960 [57] and DSP3210 [58]. We repeat their data in Table 4.1 and Table 4.2. Their data reveal what practical problems look like: the computation time is not constant and is large enough to make the LCM method inefficient. Based on their data, we construct three task graphs containing these processes as shown in Figure 4.8. The processes are allocated on four CPUs: 1 DSP3210's and 3 i960's. The allocation and priority assignment are shown in Table 4.3.

The result is given in Table 4.4 with the CPU time for running our analysis algorithm on a Sun SS20 workstation. According to the result, our algorithm has much less CPU time, more confidence, with similar quality, than an extensive simulation. We ran the extensive simulation for an interval of length equal to the LCM of the periods to compare the quality of our estimation and the efficiency of the algorithm. During the simulation, we use the lower bound value for each period and the upper bound value for each computation time. Note it is not an exhaustive simulation, where all possible values between the lower bound and upper bound should be used. Such an exhaustive simulation is too expensive to implement.

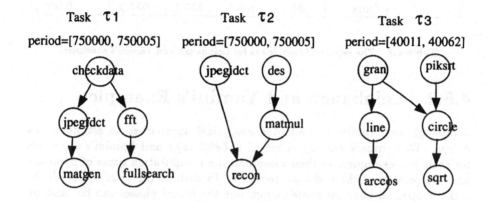

Figure 4.8 Three task graphs and their periods.

method	τ_1	τ_2	τ_3	CPU time
Our algorithm	356724	615464	29930	0.12s
Extensive simulation	355914	615004	29930	641.76s

Table 4.4 The delay estimation for the three tasks in the example created from Li and Malik's data.

Example	Method	Task 1	Task 2	Task 3	Task 4	CPU time
L & Y #1	L & Y	1116.3	1110.2	1114.3		N.A.
	Ours	1093.1	1086.0	1076.1		0.03s
L & Y #2	L & Y	958.7	845.5	863.1	911.7	N.A.
	Ours	655.4	586.5	820.5	637.3	0.03s

Table 4.5 The estimated task delays for Leinbaugh and Yamini's examples.

4.8.2   Leinbaugh and Yamini's Examples

Leinbaugh and Yamini [55] adopted analytical approaches to estimate task delays. We compare our algorithm with Leinbaugh and Yamini's algorithm for their two examples. In their examples, the computation times of processes and the periods of the tasks are constant. In each of their examples, all the tasks happen to have the same period, but the initial phases can be random. There are data dependencies between processes. Preemption is allowed. Their examples are transformed into our system model. In the first example, there are 3 tasks, 14 processes, and 3 PEs. In the second example, there are 4 tasks, 33 processes, and 3 PEs.

The results are given in Table 4.5. We got better results because they use more pessimistic assumptions. For example, they assume a process with high priority can preempt a task during the whole interval of the task's execution, even though the task only spends a portion of time on the same CPU the high-priority process is allocated to. We got better improvement (up to 32%) for the second example in Table 4.5 than the first example (up to 3%) because the computation times of highest processes in the second example (more than 20) are larger than those in the first example (about 1 or 2). When the computation times of high-priority processes are larger, the reduction of performance estimates, due to more accurate preemption calculation in our algorithm, becomes more significant.

4.8.3   D'Ambrosio and Hu's Example

Table 4.6 shows the analysis results for three designs in D'Ambrosio and Hu's example [26]. In their example, the largest period is the LCM of all the periods. There are 9 processes. There is no data dependency between processes, but each process has both a release time and a deadline. As mentioned in Section 3.1.2,

PEs	cost	D & H's simulation	Our algorithm	CPU time	Our simulation
P1, PIO	3.00	Not satisfied	Not satisfied	0.04s	Not satisfied
MC2	3.25	Satisfied	Not satisfied	0.04s	Not satisfied
MC1	3.50	Satisfied	Satisfied	0.04s	Satisfied

Table 4.6 The results about whether the deadlines of all the 9 processes are satisfied for D'Ambrosio and Hu's example.

dummy processes are added to model the release time, and there are 9 tasks. Our algorithm is compared with their simulation for three architectures. We also run an exhaustive simulation of length equal to the largest period and compare the results.

For the MC2 architecture, both our analysis algorithm and our exhaustive simulation found that the deadlines are not satisfied. But their simulation approach misjudged the design to be feasible. Less-than-exhaustive simulation eliminates the designer's confidence that the design is correct.

5

SENSITIVITY-DRIVEN CO-SYNTHESIS

5.1 OVERVIEW

This chapter describes new techniques for the co-synthesis of real-time distributed embedded systems. Embedded system synthesis is co-synthesis because the hardware and software must be designed together to meet both performance and cost goals. In contrast to traditional distributed system design, we cannot assume that the topology of the distributed system is given. In contrast with previous work on hardware-software partitioning, we do not assume a fixed template such as a one-CPU-one-ASIC configuration. Our co-synthesis algorithm [113] selects the number of PEs, the type of each PE, as well as allocating functions to PEs and scheduling their execution. The co-synthesis algorithm is based on the analytic performance estimation algorithm presented in Chapter 4. The efficiency of the performance estimation algorithm allows us to quickly explore various architectures in a large design space.

The subproblems such as scheduling, allocation, or timing analysis on a fixed-architecture distributed system are NP-hard, let alone the overall architecture synthesis problem. Gupta [30] pointed out that timing is a global property and needs to be recalculated for the entire problem; it is why incremental circuit partition algorithms such as Kernighan-Lin heuristics [46] are not suitable for hardware-software partitioning when performance is taken into consideration. The characterization and constraints for embedded systems are frequently nonlinear; it is possible to model them by either integer linear programming (ILP) or mixed integer linear programming (MILP), but may introduce a lot of extra variables and constraints, which worsen the run time of the ILP solver.

87

Most co-synthesis algorithms proposed so far belong to either of the two categories: ILP formulation or incremental optimization heuristics. The ILP approach is very slow [45, 77]. The problem model characterized by an ILP formulation is usually difficult to extend. For example, when we want to improve an ILP formulation to handle rate constraints and preemptive scheduling, the size of the formulation may increase several times, which leads to an exponential growth of run time. The majority of the co-synthesis algorithms use incremental optimization heuristics.

Our co-synthesis algorithm is an incremental optimization heuristics too. However, none of the other algorithms have provided all of the following features like ours:

- Our algorithm co-synthesizes a heterogeneous distributed systems of arbitrary topology.

- It balances multiple optimization criteria, including both delay and cost.

- It takes into consideration both hard constraints and soft constraints in either delay or cost.

- It performs preemptive scheduling under rate constraints during co-synthesis.

- Heuristic techniques can be incorporated to help jump out of local minimum during optimization.

Section 5.2 uses performance estimates and the total system cost to compute a local sensitivity of the design to allocation of processes, based on the hard and soft deadlines as well as hard and soft cost constraints. Section 5.3 proposes a new priority prediction method to reschedule processes after every change of system architecture, according to the timing criticality of processes. Section 5.4 introduces a two-stage optimization strategy and a post processing method to avoid sub-optimal designs in some cases. Based on these techniques, Section 5.5 presents a gradient-search algorithm which simultaneously design the hardware engine and the application software architecture for performance and implementation cost. Section 5.6 gives the results of experiments with the algorithm.

Communication links will not be discussed in this chapter. Chapter 6 will extend the algorithms in this chapter to synthesize communication.

5.2 SENSITIVITY ANALYSIS

Our co-synthesis algorithm uses an iterative improvement strategy. At each step, the algorithm reallocates one process from a PE to another or creates a new PE for the process. Because there are many possible reallocations, we need a measure to evaluate the candidate changes for the candidate architecture. As in most gradient-search methods, we compute a local **sensitivity**: given the current design, we estimate how much the system performance and cost will change when a single process is reallocated.

5.2.1 Nonlinearity Characterization

The hard and soft performance requirements impose non-linear constraints on the feasible solution space. For example, the improvement on a task delay over the hard deadline is significant, while the improvement on a task delay below the soft deadline is meaningless. Similar scenarios occur for the hard cost constraint and the soft cost constraint. A linear objective function, which may be used to minimize the total execution time in a non-real-time system, is not enough to optimize task delays or system costs in embedded system design.

A **design goal attribute** is either a task delay or a cost criteria. Given the ith design goal attribute, let the corresponding hard constraint or deadline be h_i, and the soft constraint or deadline be s_i. Suppose the value of the attribute in the current solution is u_i and the value will become v_i after a process reallocation and system rescheduling. The values of u_i and v_i for a task delay are estimated by the performance analysis algorithm in Chapter 4 with the scheduling method discussed in Section 5.3. The values of u_i and v_i for a cost criteria are the sum of the component costs for all the PEs in the system.

The *displacement vector* **D** represents the change of design goal attributes after a process reallocation and system rescheduling. Define the ith component of the displacement vector to be

$$D_i = \frac{1}{h_i} \int_{u_i}^{v_i} W(x)dx$$

where $W(x)$ is a weight function given in Figure 5.1. In other words, D_i represents the amount of the change $v_i - u_i$, but we give higher weight (penalty)

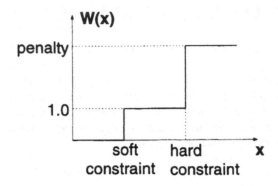

Figure 5.1 The weight function which emphasizes the change above the hard constraint and ignores the change below the soft constraint.

for the portion of change above the hard constraint, and no credit (zero) for the portion below the soft constraint. It is divided by h_i for normalization—the tighter the hard constraint is, the higher the weight on the change. In our heuristics, we set the value of the penalty in Figure 5.1 to be the number of design goal attributes plus one. The displacement vector characterizes the nonlinear design goal under the hard and soft constraints.

5.2.2   Multi-Dimensional Goal Balancing

Embedded system performance is not characterized by a single number, since each task can have its own deadline to meet. In non-real-time systems, usually the main objective is to minimize the total execution time of all the tasks; we can evaluate the system by the magnitude of a single number—the total execution time [89]. In real-time systems, the satisfaction of the deadlines is more important than shorter delays. Because each task has its individual deadline to meet, the problem is represented by multiple delay values instead of a single number. Unlike design problems on a fixed architecture, we need to also take into consideration the change in other cost criteria such as price, area, or power consumption, since the underlying hardware may change.

The *target vector* **T** describes the distance between the current solution and the attribute values of the ideal system. Let the ith component of the target vector be

$$T_i = (s_i - u_i)/h_i \tag{5.1}$$

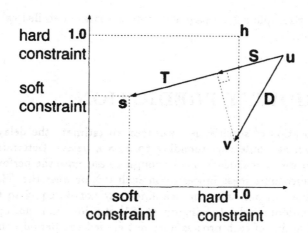

Figure 5.2 The illustration for the sensitivity vector S, which is the orthogonal projection of the displacement vector on the target vector. The current solution is at u, and the next solution is at v. The sensitivity is the magnitude of S.

As shown in Figure 5.2, because the final goal is the soft constraint, the vector from the current position to the soft constraint is the direction in which we want to move. We also normalize each component of the vector by the hard constraint value h_i.

The sensitivity S is the magnitude of the projection of **D** on **T**, that is:

$$S = \frac{\mathbf{D} \cdot \mathbf{T}}{\|\mathbf{T}\|}$$

The sensitivity tells how much closer we move the design towards the target, as illustrated in Figure 5.2.

The target vector automatically balances weights among different design goals, according to the current system architecture. The attributes with tighter hard constraints and the attributes whose current values are farther from soft constraints are weighted more heavily.

A positive sensitivity implies an improvement, while a negative value means the result may become worse. We only adopt a system change with a positive sensitivity. A larger sensitivity implies a better improvement on the system architecture. Example 5.2 on page 97 demonstrates a detailed calculation of the sensitivities.

5.3   PRIORITY PREDICTION

In the computation of sensitivities, we need to estimate the delay values u_i and v_i for each attribute corresponding to a task delay. Determining where and which process to reallocate is not enough to estimate the performance; we also need to reschedule these processes on each PE, because the delays depend on the schedule. In our case, the deadline may not be equal to the period, so the rate-monotonic priority assignment [61] of priorities is not optimal. We assume the deadline of each process does not exceed the period—this suggests a task with a small period should be partitioned into processes with a finer granularity to facilitate task pipelining, though we do not discuss partitioning in this work. If the deadline is smaller than or equal to the period for a process, the inverse-deadline priority assignment [56] is optimal for one processor. However, in our model the deadline is specified end-to-end for a whole task, not for individual processes. We develop a heuristic to use the inverse-deadline priority assignment.

5.3.1   Fractional Deadline

As illustrated in Figure 5.3, we define the **fractional deadline** of a process—the portion of the task deadline which a particular process must meet—as follows. Algorithm LatestTimes in Figure 4.2 on page 65 calculates the *latest request time* and *latest finish time* relative to the start of a task for each process. Assign each process a weight equal to its latest finish time minus its latest request time. For each CPU R, temporarily assign weight zero to the set of processes $\mathcal{J}^R$ allocated on R in the task. Then apply the longest-path algorithm backwards from the end of the task. The **latest required time** of each process in $\mathcal{J}^R$ is the hard deadline of the task minus the longest path weight of the process from the end of the task. The calculation of latest required times is similar to the technique in as-late-as-possible (ALAP) scheduling of high-level synthesis [65]. The latest required time is the time before which the process must finish in order not to violate the task deadline when the processes allocated to other PEs run in their worst case. The fractional deadline d_i of each process $P_i \in \mathcal{J}^R$ is its latest required time minus its latest request time. We can then

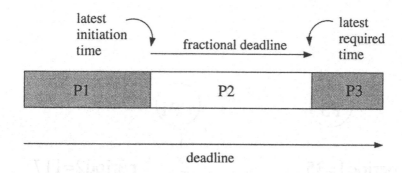

Figure 5.3 The fractional deadline of a process in a task graph.

Process	P_1	P_2	P_3	P_4	P_5
latest request time	0	15	35	55	0
latest finish time	15	35	55	81	17
latest required time	50	64	64	90	34
fractional deadline	50	49	29	35	34

Table 5.1 The fractional deadline calculation for the processes in Figure 5.4.

order the priority by d_i—the shorter the fractional deadline, the higher the priority.

Example 5.1: For the example in Figure 5.4, suppose in the current design, P_2 and P_3 are allocated to a PE of type X, P_1 and P_4 are allocated to a PE of type Y, and P_5 is allocated to a PE of type Z. The fractional deadlines of processes are listed in Table 5.1. If we move P_5 from Z to X, the three processes P_2, P_3, and P_5 will share the same PE. According to the fractional deadlines, their priorities should be ordered as $P_3 > P_5 > P_2$. The task deadlines are satisfied under this schedule. Among six possible schedules for three processes, this is the only feasible schedule for the example. ■

Intuitively, the way we calculate fractional deadlines approximates the deadline driven scheduling algorithm [61], which dynamically assigns the highest priority to a process with a deadline nearest to its current request.

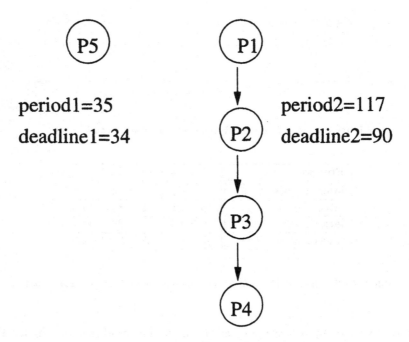

two task graphs

PE type	Computation time				
	P_1	P_2	P_3	P_4	P_5
X	–	20	20	–	10
Y	15	–	–	26	–
Z	–	–	–	–	17

process characteristics

Figure 5.4   An example of fractional deadlines.

5.3.2 Prediction Scheduling

Unfortunately, when we reallocate a process, the delay information about latest initiation times and latest termination times will change. We should use the new delay information to decide the process deadlines for priority assignment, but before we schedule the priority, we cannot know the new delay values. We solve this problem by using the delay information in the current solution before the reallocation to *predict* the fractional deadlines of processes and assign priority in the next solution, and use the new schedule to compute the new delay information. The new delay information is used to decide priority assignment in the next step. Since we only move one process at a time, the change in the delay is usually not global, making this an acceptable approximation.

5.4 OPTIMIZATION BY STAGES

The sensitivity only gives a local measure of the design optimality. It is very possible that our incremental optimization approach will be trapped into a local optimal solution. This is unavoidable for a heuristic algorithm, since searching for a global optimal solution is impractical for such an NP-hard problem. Nevertheless, we still want to increase the chance of jumping off a local optimal solution.

Simulated annealing is a common approach to deal with such situations [27]. However, simulated annealing requires a long run time and the quality of the solution is sensitive to the cooling schedule [44]. Given multiple nonlinear design goals in real-time embedded systems, a good objective function for simulated annealing may not be available.

Instead of simulated annealing, we adopt efficient heuristics which help avoid local optimal solution in some cases. Unlike simulated annealing which is a general approach, the heuristics in this section is customized for our co-synthesis problem. The heuristics divide the co-synthesis algorithm into stages with different criterion for optimization.

5.4.1 Idle-PE Elimination

In the computation of sensitivities, we need to know the total system cost. The system cost is usually the sum of the cost of each individual component.

Consider the situation where two processes are on a PE R, and it is feasible to move these two processes to another PE, remove R, and reduce the cost. Since we are allowed to move only one process at a time, when the first process on R is reallocated, cost is not reduced immediately and such a movement may induce additional delay caused by a higher load on another PE. On the other hand, if R cannot be removed for a feasible solution, it might be desirable to move some processes from a highly-utilized PE to R to increase the performance. To solve the dilemma and maximize the satisfaction of goals during both the early and late stages of optimization, we use different criteria at different stages of synthesis:

- **idle-PE-elimination stage:** If R is the least-utilized PE, add the product of the cost and the processor utilization of R to the total system cost. In other words, the calculation of cost is modified by taking into account the utilization of the least-utilized PE. When we move one of the processes from the least-utilized PE to another PE, we can immediately feel that the system cost is reduced, which increases the possibility to accept the first move and then take the other processes away during the next steps and remove R.

- **load-balancing stage:** After we have removed as many PEs as we can, we calculate the cost function as usual without considering the least PE utilization, concentrating on balancing PE utilization to increase the performance whenever it is possible.

The idle-PE-elimination criterion helps the solution jump off local minima; the load-balancing criterion brings the solution back to a minimum if a better solution cannot be found.

The processor utilization is defined in formula (2.1) on page 15. The least-utilized PE is the PE whose processor utilization is smallest in the current system. Let the least-utilized PE in the current solution be R_1 and its processor utilization be U_{least}^{before}. Let the least-utilized PE in the next solution after process reallocation be R_2 and its processor utilization be U_{least}^{after}. It is possible that $R_1 = R_2$. For a certain cost attribute i, during the idle-PE-elimination stage, the value of v_i, as defined in Section 5.2.1, should be modified by

$$v_i^{idle-PE-elimination} = v_i + (U_{least}^{after}) \times R_2.cost_i - (U_{least}^{before}) \times R_1.cost_i \quad (5.2)$$

The value of $v_i^{idle-PE-elimination}$ will replace v_i in the sensitivity calculation.

Example 5.2: We are given the following system specification:

- Four independent processes P_1, P_2, P_3, and P_4, whose periods are $p_1=50$, $p_2=51$, $p_3 = 52$, and $p_4 =53$.

- The hard deadline of each process equals the period. The soft deadlines are all zeros.

- For a CPU of type X with cost 6, the computation times of the processes on type X are $c_1 = c_2 = c_3 = c_4 = 10$.

- The total system cost has a hard constraint of 10 and a soft constraint of 0.

Suppose after some optimization steps, the system configuration is as follows:

- The hardware engine contains two CPUs, R_1 and R_2, both of type X.

- Processes P_1 and P_2 are allocated to PE R_1, while P_3 and P_4 are allocated to R_2.

- Rate-monotonic scheduling is applied to schedule processes on the CPUs.

Given such a system configuration, we may further improve the embedded system architecture by moving a process from one PE to another. Let the worst-case response times of the four processes be the first four design attributes, and the total system cost be the fifth design attribute. Then

$$h = (50, 51, 52, 53, 10)$$

$$s = (0, 0, 0, 0, 0)$$

and the current solution is

$$u = (10, 20, 10, 10, 12)$$

By formula (5.1), the target vector is

$$T = (-0.2, -0.392, -0.192, -0.377, -1.2)$$

There are four possible reallocation of processes; if no idle-PE elimination technique is applied, their sensitivity values are calculated as follows:

- Move P_1 from R_1 to R_2:

    $$v = (10, 10, 20, 20, 0)$$

    $$D = (0, -0.196, 0.192, 0.189, 0)$$

    $$S = -0.0232$$

- Move P_2 from R_1 to R_2:

    $$v = (10, 10, 20, 20, 0)$$

    $$D = (0, -0.196, 0.192, 0.189, 0)$$

    $$S = -0.0232$$

- Move P_3 from R_2 to R_1:

    $$v = (10, 20, 30, 10, 0)$$

    $$D = (0, 0, 0.385, -0.189, 0)$$

    $$S = -0.00205$$

- Move P_4 from R_2 to R_1:

    $$v = (10, 20, 10, 30, 0)$$

    $$D = (0, 0, 0, 0.187, 0)$$

    $$S = -0.0529$$

Since all the sensitivity values are negative, no movement can be adopted, and the current system configuration becomes the final embedded system architecture, where the hard cost constraint is violated and no feasible design is found.

On the other hand, if we use formula (5.2) to modify the cost calculation for idle-PE elimination, the sensitivities are

- Move P_1 from R_1 to R_2:

    $v = (10, 10, 20, 20, 10.8)$

    $D = (0, -0.196, 0.192, 0.189, -0.72)$

    $S = 0.618409$

- Move P_2 from R_1 to R_2:

    $v = (10, 10, 20, 20, 10.82)$

    $D = (0, -0.196, 0.192, 0.189, -0.706)$

    $S = 0.605827$

- Move P_3 from R_2 to R_1:

    $v = (10, 20, 30, 10, 10.85)$

    $D = (0, 0, 0.385, -0.189, -0.692)$

    $S = 0.614913$

- Move P_4 from R_2 to R_1:

    $v = (10, 20, 10, 30, 10.87)$

    $D = (0, 0, 0, 0.187, -0.679)$

    $S = 0.552449$

The sensitivities become positive, and the first move which takes P_1 from R_1 to R_2 will be taken. In the next step of optimization, P_2 will be easily moved from R_1 to R_2, R_1 can be removed, and only one PE remains. By the idle-PE elimination technique, a feasible system with a cost of only 6 can be obtained. ∎

Load balancing stage is necessary after the idle-PE-elimination stage. The importance of load balancing stage can be explained by the example in Table 5.3.

5.4.2   PE Downgrading

The types of PEs are not known in advance during distributed system co-synthesis. Since we cannot determine how many processes will be allocated to a PE when the PE is created, a PE type larger than necessary may be chosen at the beginning.

PE downgrading is a post-processing procedure performed after all optimization steps have finished and no process reallocation will take place. In this final stage, we have known which and how many processes are allocated to each PE. Replace a PE R_i of PE type X with another PE type Y $(X \neq Y)$, whenever all of the following conditions are satisfied:

- PE type Y has a lower cost than PE type X.

- PE type Y can implement all the processes on R_i such that no hard deadlines are violated.

- The sensitivity is positive for such a change of the embedded system architecture.

- The sensitivity value is the largest among all possible replacements which satisfy the above conditions.

This PE downgrading procedure is repeated until no replacement is feasible.

5.5 THE CO-SYNTHESIS ALGORITHM

Based on the methodology described in the previous sections, our complete synthesis algorithm consists of the following steps:

1. Create an initial solution by assigning only one process to each PE. The PE type with highest performance-to-cost ratio is chosen for each process.

2. Call optimize(idle-PE-elimination), an iterative optimization procedure in the idle-PE elimination stage. Procedure optimize(idle-PE-elimination criterion) will be described in Section 5.5.2 in detail.

3. Call optimize(load-balancing), an iterative optimization procedure in the load-balancing stage. Procedure optimize(idle-PE-elimination criterion) will be described in Section 5.5.2 in detail.

4. Perform PE downgrading, as described in Section 5.4.2, to replace PEs which are poorly utilized.

5.5.1 Initial Solution

We choose a PE for each process in the initial solution. If a process computation time on a certain PE type is longer than the period of the task the process belongs to, that PE type will not be considered at all. Suppose process P_i has a computation time c_i^X on PE type X, and the cost of PE type X is $cost_X$. Choose a type X for P_i such that $c_i^X \times cost_X$ is the largest. A better cost-performance ratio for each process in the initial solution gets higher probability to reach a better final solution.

The priority prediction technique described in Section 5.3 cannot be applied to the initial solution, because there is no previous solution providing delay information for prediction. However, since there is only one process on each CPU in our initial solution, there is no need for scheduling. All the following solutions can depend on the delay information in the previous one for scheduling.

Due to the *process cohesion* relation described in Section 3.2.2 on page 50, we may still need to put more than one process on the same PE. In that case, these processes are scheduled by rate-monotonic priority assignments in the initial solution.

5.5.2   Iterative Optimization Procedures

Our co-synthesis algorithm uses two optimization procedures, optimize(idle-PE-elimination) and optimize(load-balancing). These two procedures are similar, but are parameterized by the optimization criterion to be used.

These two procedure operate as follows:

1. Compute the sensitivity for each possible movement of a process from a PE to another, as described in Section 5.2. If there are $|P|$ processes and $|R|$ PEs in the current solution, there are at most $|P| \times (|R| - 1)$ possible movements.

2. Ignore the movements which belong to one of the following categories:

   - Disallow a process from moving back to the same PE or the same type of PE if it is moved away from that PE in a previous iteration. This avoids repeating a certain system configuration.

   - Do not consider the movements separating two processes which should be on the same PE according to the *process cohesion* relation defined in Section 3.2.2 on page 50. Do not consider the movements putting two processes together on the same PE and violating a *process exclusion* relation defined in Section 3.2.2.

   - Exclude those movements whose sensitivities are negative. A negative value of sensitivity implies that the movement may make the design worse.

3. When no movement remains feasible after step 2, create a new PE of a certain type and move a process to the new PE. Compute the sensitivities for each possible PE type and each process. If there are $|P|$ processes and $|R_t|$ different PE types, there are at most $|P| \times |R_t|$ possible creations. Exclude the choices whose sensitivities are negative. If there is still no feasible movement, return.

4. Among the remaining movements or creations, choose the one with highest sensitivity. Make such a movement and reschedule each PE using priority prediction mentioned in Section 5.3, according to the information in the current solution about how critical each process is. When two processes have process cohesion relation, they must be moved together.

5. Repeat steps 1–4 until no movement is feasible.

Example	Problem size			The result		CPU
	#task	#process	#PE type	#PE	cost	time
ex1	2	6	3	2	$1200	0.72s
dambrosio-h	9	9	11	1	$3.50	31.2s
prakash-p-1	1	9	3	see Table 5.4		
prakash-p-2	2	13	3	3	$12	119.78s
random-1	6	50	8	13	$281	10252.5s
random-2	8	60	12	11	$637	21978.9s

Table 5.2 The problem size, the final result, and the CPU time of running our algorithm for each example.

5.5.3   Analysis

In step 2 of procedures optimize($\cdot$), we disallow a processor from being moved back to the same PE or the same type of PE it is moved away from in a previous iteration. Under this condition, the number of iterations in each call to optimize(idle-PE-elimination) or optimize(load-balancing) is bounded by $|P|^2 \times |R_t|$, where $|P|$ is the number of processes and $|R_t|$ is the number of PE types. Because each process can be allocated to at most $|R_t|$ different PE types, we can have at most $|P|$ PEs for each PE type. In practice, the algorithm converges to a solution in many fewer steps.

5.6   EXPERIMENTAL RESULTS

We implemented our algorithm in C++ and performed experiments on several examples. All experiments were performed on a Sun SS20 Sparc Workstation. The results of all our experiments are summarized in Table 5.2.

5.6.1   A Step-by-Step Example

The first example, *ex1*, is a small example shown in Figure 5.5. We assume that the designer wants to reduce the delay and the cost as much as possible, so the soft cost constraint and soft deadlines are set to zero to encourage optimization

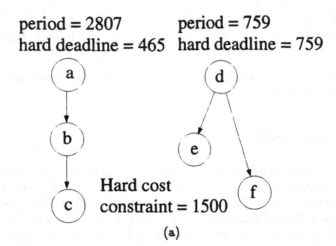

period = 2807   period = 759
hard deadline = 465   hard deadline = 759

Hard cost
constraint = 1500

(a)

PE type	Cost	Computation time					
		a	b	c	d	e	f
X	$800	179	100	95	213	367	75
Y	$500	204	173	124	372	394	84
Z	$400	210	193	130	399	494	91

(b)

Figure 5.5 A small example *ex1*. (a) The task graphs for two tasks, their periods and deadlines. (b) The computation time of each process and the cost on each type of PE.

Step	embedded system architecture	hard deadlines	cost
	Initial Solution		
1	(Y: e) (Z: a) (Z: b) (Z: c) (Z: d) (Z: f)	unsatisfied	$2500
	Stage 1: idle-PE elimination		
2	(X: b) (Y: e) (Z: a) (Z: c) (Z: d) (Z: f)	unsatisfied	$2900
3	(X: b d) (Y: e) (Z: a) (Z: c) (Z: f)	satisfied	$2500
4	(X: b e d) (Z: a) (Z: c) (Z: f)	satisfied	$2000
5	(X: b e d) (Z: c a) (Z: f)	satisfied	$1600
6	(X: b e d) (Z: c a f)	satisfied	$1200
7	(X: b e d f) (Z: c a)	satisfied	$1200
	Stage 2: load balancing		
8	(X: b e d) (Z: c a f)	satisfied	$1200
	Stage 3: PE downgrading		
9	(X: b e d) (Z: c a f)	satisfied	$1200

Table 5.3 The iterative optimization for *ex1*. Each processor in the system is shown by its type, followed by a colon, and then the processes allocated to it are listed from the highest priority to the lowest.

whenever it is possible. The embedded system architecture, total cost and the satisfaction of real-time deadlines after each iterative step are given in Table 5.3.

In the initial solution, there is only one process on each PE. The PE type is chosen for highest performance-to-cost ratio. In other words, if the computation time of a process is c, the PE type T will chosen so that $T.cost \times c$ is largest.

It took 6 steps in the idle-PE elimination stage. In the last step, process f is taken from Z to X to increase the possibility to remove Z. However, it is impossible to remove Z and satisfy the hard deadlines, so in the load-balancing stage, process f is moved back to Z. All the PEs have been highly utilized, so PE downgrading stage does not change the design.

5.6.2 D'Ambrosio and Hu's Example

The second example is created by D'Ambrosio and Hu [26]. Their algorithm is reviewed in Section 2.5.4 on page 35. The PE downgrading stage is important for this example because some larger CPUs have better performance-to-cost ratio and are chosen first in the initial solution.

Design	Hard cost constraint	Prakash & Parker's		Our results	
		delay	CPU time	delay	CPU time
1	$15	5	62.20min	6	9.43sec
2	$12	6	445.17min	7	11.80sec
3	$8	7	538.67min	9	8.15sec
4	$7	8	75.18min	9	8.13sec
5	$5	15	6416.87min	-	7.68sec

Table 5.4 The results for Prakash and Parker's examples. Prakash and Parker's run times were measured on a Solbourne 5e/900 (similar to Sun SPARC 4/490). Our run times were measured on a Sun Sparc20 Workstation. For Design 5, our algorithm cannot give a solution satisfying the hard cost constraint.

Our result, whose cost is 3.50, is the second best shown in their results. However, as shown in Section 4.8.3 on page 84, by exhaustive simulation, we found their best design with the minimum cost of 3.25 is actually infeasible. D'Ambrosio and Hu used simulation to verify whether a system configuration is feasible, but simulation does not guarantee the deadlines are satisfied. Our result is actually the best feasible design achievable.

5.6.3 Prakash and Parker's Examples

Our third example is the second example of Prakash and Parker [77]; their algorithm is summarized in Section 2.5.1 on page 34. We use their assumption on communication cost and delay, and only point-to-point interconnections are allowed. (The algorithm which handles more sophisticated communication topology will be discussed in the next chapter.) The problem size is shown in the row labeled *prakash-p-1* in Table 5.2. We compare our results with theirs in Table 5.4 for five designs with different hard cost constraints. Our result for the task delay is worse than, but close to their result. Their integer linear programming approach is optimal but takes hours on this example, while ours takes only seconds. Even though after we take into consideration the different speed of workstation, our algorithm is still much faster.

Our algorithm can co-synthesize from multiple disjoint task graphs, but Prakash and Parker's approach can only handle a single task graph. We also combine Prakash and Parker's two examples together and assign the periods as well as

Task	worst-case delay
τ_1	15606
τ_2	25506
τ_3	24991
τ_4	57719
τ_5	29800
τ_6	8781

Table 5.5 Task worst-case delays in the result of *random-1*.

the hard deadlines of 7 and 15 to the two tasks. The soft deadlines are assumed to be zero. The results are given under the row labeled *prakash-p-2* in Table 5.2.

5.6.4 Random Large Examples

In order to test the performance of our algorithm on larger examples, we randomly generate two examples, as shown in the rows labeled *random-1* and *random-2* in Table 5.2. The task delays in the results are given in Table 5.5 and Table 5.10. The system specification for *random-1* is given in Table 5.6, Table 5.7, Table 5.8, and Table 5.9. The system specification for *random-2* is given in Table 5.11, Table 5.12, Table 5.13, Table 5.14, and Table 5.15.

It may take a few hours for our algorithm to handle an example with more than 50 processes. The efficiency of co-synthesis algorithms is an important issue. However, our algorithm is still much faster that approaches such as integer linear programming or simulated annealing.

5.6.5 Comparison

Compared with existing techniques, the chapter results are sumarized as follows:

- The quality of our results is worse than mathematical programming, but close to other heuristics.

Task	period	soft deadline	hard deadline
τ_1	153074	0	30614
τ_2	239234	0	47846
τ_3	153830	0	30766
τ_4	441464	0	88292
τ_5	261481	0	52296
τ_6	128004	0	25600

Hard cost constraint = 8200
Soft cost constraint = 5

Table 5.6   Task periods and deadlines in example *random-1*.

- The run time efficiency of our algorithm is much better than mathematical programming and exhaustive search.

- The capability of our algorithm is better than mathematical programming and exhaustive search. We can handle multiple task graphs and preemptive scheduling, etc. However, we cannot have examples to compare results if existing techniques cannot handle such problems.

Task	data dependency edges
τ_1	P1/P6, P3/P4, P2/P3, P2/P4, P1/P5, P4/P6, P3/P5, START/P1, START/P2, P5/END, P6/END
τ_2	P11/P13, P7/P14, P8/P15, P9/P13, P8/P11, P8/P14, P12/P14, P8/P16, P11/P15, P8/P13, P8/P9, P12/P13, P7/P11, P7/P9, P11/P14, P8/P10, P11/P14, P7/P15, P12/P13, START/P7, START/P8, P10/END, START/P12, P13/END, P14/END, P15/END, P16/END
τ_3	P17/P18, P17/P20, P17/P19, P20/P21, P19/P20, P17/P21, START/P17, P18/END, P21/END
τ_4	P28/P32, P27/P34, P25/P30, P23/P30, P31/P35, P23/P34, P23/P28, P22/P34, P23/P32, P27/P32, P27/P35, P26/P35, P25/P34, P26/P29, P22/P28, P27/P31, P23/P29, P26/P31, P22/P24, P29/P35, P28/P34, P31/P32, P26/P34, P29/P34, P22/P35, P23/P26, P29/P35, START/P22, START/P23, P24/END, START/P25, START/P27, P30/END, P32/END, START/P33, P33/END, P34/END, P35/END
τ_5	P44/P45, P36/P43, P39/P43, P37/P42, P42/P44, P37/P39, P36/P45, P40/P45, P40/P44, P40/P43, P36/P42, START/P36, START/P37, START/P38, P38/END, START/P40, START/P41, P41/END, P43/END, P45/END
τ_6	P47/P48, P49/P50, P46/P50, P48/P49, P48/P49, P49/P50, P48/P49, P46/P47, P46/P48, START/P46, P50/END

Table 5.7 Task graphs in example *random-1*. P1/P2 represents a data dependency edge from process P1 to process P2.

PE type	process computation time
Y1 (cost = 155)	P1=207, P2=113, P3=158, P5=295, P6=88, P7=200, P8=97, P9=173, P10=79, P11=66, P12=132, P13=96, P14=142, P15=284, P16=204, P17=212, P18=179, P20=267, P21=267, P23=168, P25=144, P26=287, P27=137, P29=159, P32=281, P34=196, P36=222, P37=180, P38=124, P39=99, P41=142, P42=128, P43=310, P44=90, P45=85, P46=233, P47=261, P48=162, P49=282, P50=270
Y2 (cost = 122)	P1=196, P4=185, P5=415, P6=100, P10=98, P11=87, P12=183, P13=188, P15=324, P16=279, P17=282, P18=244, P19=313, P20=235, P21=236, P22=135, P23=235, P26=297, P27=187, P28=168, P29=250, P30=269, P31=267, P32=307, P33=262, P34=277, P35=293, P37=160, P38=141, P39=158, P40=335, P41=194, P42=248, P43=290, P44=100, P45=122, P46=415, P47=279, P48=178, P49=240, P50=221
Y3 (cost = 46)	P1=511, P2=394, P3=493, P5=682, P6=396, P7=692, P8=332, P9=672, P10=290, P11=257, P12=444, P13=393, P15=1110, P18=638, P20=750, P21=953, P22=387, P23=585, P24=480, P25=487, P27=510, P28=385, P29=783, P32=776, P33=814, P34=701, P35=846, P36=478, P37=386, P38=469, P39=303, P40=923, P41=480, P42=524, P43=852, P46=778, P47=750, P48=597, P50=806
Y4 (cost = 5)	P1=4916, P2=2665, P3=4198, P4=4238, P5=8743, P7=5296, P8=3571, P9=4600, P10=2784, P11=2960, P12=4644, P13=4733, P14=4094, P16=5606, P17=5977, P18=7179, P19=8288, P20=8778, P21=8703, P22=3691, P23=6958, P24=4722, P25=4369, P26=7984, P27=5325, P28=5688, P29=4879, P30=7808, P31=9383, P32=6628, P34=5958, P35=7897, P36=7182, P37=5206, P38=4830, P39=4041, P40=8694, P41=5432, P43=8730, P44=3282, P45=3432, P47=5853, P49=5850, P50=5999

Table 5.8 PE types, their cost, and process computation time in example *random-1*. (To be continued by Table 5.9.)

PE type	process computation time
Y5 (cost = 164)	P1=123, P2=106, P3=125, P4=156, P5=203, P8=112, P9=153, P10=77, P11=76, P12=133, P13=121, P14=119, P16=192, P17=228, P18=172, P19=246, P21=193, P23=170, P24=140, P25=103, P26=266, P28=111, P30=242, P31=295, P33=218, P34=150, P35=178, P36=178, P37=167, P38=138, P39=86, P41=131, P43=310, P44=72, P45=93, P46=221, P47=168, P48=124, P50=159
Y6 (cost = 120)	P3=154, P5=309, P7=252, P8=150, P9=197, P11=116, P12=226, P13=171, P15=273, P16=175, P19=246, P20=245, P21=340, P22=121, P23=297, P24=156, P25=188, P26=353, P27=228, P28=212, P29=213, P30=255, P32=286, P33=336, P34=289, P35=260, P37=214, P38=180, P39=159, P40=241, P41=166, P42=249, P44=120, P45=148, P46=356, P47=285, P48=223, P49=287, P50=227
Y7 (cost = 24)	P1=896, P2=799, P3=930, P4=1236, P5=1761, P7=1207, P8=843, P9=1352, P10=474, P11=505, P12=933, P15=2050, P17=1180, P18=1514, P19=1533, P20=1187, P21=1326, P22=587, P23=1246, P24=958, P25=1019, P26=1776, P27=1112, P29=938, P30=1509, P31=1462, P32=1337, P33=1779, P35=1669, P36=951, P37=1165, P38=797, P39=752, P40=1394, P41=1005, P42=911, P43=2014, P45=747, P47=1753, P48=1204, P49=1304, P50=1236
Y8 (cost = 129)	P1=175, P2=97, P4=222, P6=93, P7=245, P8=128, P9=153, P10=96, P11=91, P12=137, P13=123, P15=282, P17=243, P18=193, P19=322, P20=264, P21=274, P22=112, P23=253, P25=188, P26=329, P27=224, P28=147, P29=286, P31=323, P32=267, P33=246, P35=264, P36=259, P37=157, P38=134, P39=127, P41=153, P42=177, P43=278, P44=91, P46=310, P48=164, P49=296, P50=240

Table 5.9 (Continued from Table 5.8.) PE types, their cost, and process computation time in example *random-1*.

Task	worst-case delay
τ_1	2670
τ_2	2978
τ_3	9355
τ_4	3715
τ_5	10404
τ_6	3866
τ_7	3275
τ_8	16498

Table 5.10 Task worst-case delays in the result of *random-2*.

Task	period	soft deadline	hard deadline
τ_1	246307	0	30788
τ_2	146001	0	18250
τ_3	293088	0	36636
τ_4	206186	0	25773
τ_5	299128	0	37391
τ_6	151503	0	18937
τ_7	142354	0	17794
τ_8	132487	0	16560

Hard cost constraint $= 11520$
Soft cost constraint $= 8$

Table 5.11 Task periods and deadlines in example *random-2*.

Task	data dependency edges
τ_1	P5/P11, P3/P9, P8/P10, P8/P9, P3/P4, P5/P6, P2/P3, P1/P8, P1/P2, P5/P10, P6/P7, P5/P7, P1/P5, P3/P10, P2/P9, P3/P6, P1/P10, P4/P5, P9/P10, P2/P5, START/P1, P7/END, P10/END, P11/END
τ_2	P13/P14, P12/P13, P14/P16, P13/P15, P14/P16, P12/P14, P12/P15, P14/P16, START/P12, P15/END, P16/END
τ_3	P23/P26, P19/P20, P21/P25, P21/P26, P17/P19, P17/P26, P22/P26, P19/P23, P18/P26, P25/P26, P17/P21, P20/P21, P22/P24, P24/P26, P20/P26, START/P17, START/P18, START/P22, P26/END
τ_4	P30/P31, P29/P30, P28/P29, P28/P31, P29/P31, P28/P30, P27/P29, P31/P32, START/P27, START/P28, P32/END
τ_5	P40/P42, P33/P40, P34/P38, P39/P43, P39/P41, P37/P43, P37/P42, P33/P35, P35/P37, P40/P41, P36/P40, P33/P36, START/P33, START/P34, P38/END, START/P39, P41/END, P42/END, P43/END
τ_6	P46/P47, P44/P46, P46/P48, P44/P48, P45/P47, P45/P48, START/P44, START/P45, P47/END, P48/END
τ_7	P50/P53, P49/P51, P50/P52, P49/P50, P51/P53, START/P49, P52/END, P53/END
τ_8	P55/P56, P56/P59, P54/P58, P54/P59, P55/P59, P56/P60, P54/P56, P54/P55, START/P54, START/P57, P57/END, P58/END, P59/END, P60/END

Table 5.12 Task graphs in example *random-2*. P1/P2 represents a data dependency edge from process P1 to process P2.

PE type	process computation time
Y1 (cost = 81)	P1=327, P2=151, P3=241, P4=289, P5=572, P6=218, P7=361, P9=277, P10=160, P12=274, P14=245, P15=442, P16=271, P18=303, P21=481, P23=450, P25=234, P26=540, P27=384, P28=228, P30=387, P31=490, P32=529, P33=374, P34=354, P35=326, P36=432, P37=284, P38=235, P39=183, P40=520, P43=595, P44=207, P45=209, P49=494, P50=349, P51=603, P52=108, P54=401, P55=266, P56=384, P57=139, P58=519, P59=195, P60=171
Y2 (cost = 185)	P1=168, P2=68, P5=261, P6=70, P7=213, P8=107, P9=168, P10=61, P11=84, P12=94, P13=79, P14=128, P16=168, P17=190, P18=134, P19=215, P20=193, P21=245, P22=88, P23=174, P24=132, P25=93, P26=267, P27=138, P28=112, P29=160, P30=194, P31=262, P32=183, P34=133, P35=225, P36=158, P37=94, P38=133, P39=82, P40=175, P41=106, P42=152, P44=83, P47=142, P49=186, P50=208, P51=254, P52=65, P53=122, P54=150, P56=120, P57=74, P58=260
Y3 (cost = 153)	P1=134, P2=124, P3=102, P4=188, P5=233, P6=80, P7=232, P8=114, P9=182, P10=88, P11=97, P12=166, P15=339, P16=223, P17=218, P19=200, P20=211, P21=182, P22=87, P23=155, P24=122, P26=275, P27=191, P28=170, P29=157, P30=281, P31=299, P32=291, P33=223, P35=245, P36=164, P37=157, P38=134, P39=114, P40=271, P41=131, P42=151, P43=224, P44=103, P45=94, P46=273, P47=216, P48=119, P49=186, P50=183, P51=255, P53=163, P54=246, P55=128, P56=162, P57=88, P58=248, P59=100, P60=83
Y4 (cost = 96)	P2=154, P3=166, P4=247, P6=135, P8=166, P9=230, P10=175, P11=163, P12=239, P13=192, P14=224, P17=286, P18=330, P19=402, P20=360, P21=411, P22=167, P23=366, P25=219, P26=395, P27=335, P28=283, P30=423, P32=460, P33=403, P34=278, P35=396, P37=253, P39=174, P40=333, P42=323, P43=490, P44=158, P45=177, P46=469, P47=291, P48=236, P50=363, P52=131, P53=236, P54=412, P55=203, P56=377, P57=140, P59=160

Table 5.13   PE types, their cost, and process computation time in example *random-2*. (To be continued by Table 5.14.)

PE type	process computation time
Y5 (cost = 72)	P2=249, P4=386, P6=165, P7=402, P8=226, P10=172, P11=145, P12=288, P14=367, P15=446, P18=446, P19=385, P20=549, P21=491, P22=203, P23=333, P24=270, P25=223, P26=640, P27=445, P28=351, P29=500, P31=443, P32=647, P33=621, P34=551, P36=349, P37=282, P38=265, P39=191, P40=551, P41=326, P43=708, P44=189, P45=231, P46=474, P47=440, P48=367, P49=404, P50=457, P51=492, P52=150, P53=310, P54=377, P56=463, P57=146, P58=677, P59=291, P60=156
Y6 (cost = 61)	P1=489, P4=382, P6=203, P7=653, P8=247, P9=406, P10=219, P11=172, P12=464, P13=302, P14=402, P15=584, P16=512, P18=588, P19=488, P20=723, P21=462, P22=211, P24=281, P25=435, P26=520, P27=533, P28=345, P29=550, P30=642, P31=806, P32=552, P34=434, P36=452, P37=448, P38=376, P39=256, P40=699, P41=367, P43=809, P45=250, P46=708, P47=605, P48=351, P49=550, P50=490, P51=711, P52=223, P54=513, P56=383, P57=162, P58=610, P59=385, P60=183
Y7 (cost = 50)	P1=505, P2=327, P4=423, P5=682, P6=353, P7=560, P8=270, P10=259, P11=304, P12=423, P14=477, P15=805, P16=548, P17=828, P18=599, P20=790, P21=923, P22=263, P23=534, P25=382, P27=613, P28=530, P29=515, P30=851, P31=898, P32=660, P33=896, P34=534, P35=547, P37=483, P38=523, P39=349, P40=733, P41=379, P42=611, P43=784, P45=411, P46=656, P47=776, P49=823, P50=748, P52=176, P53=308, P54=827, P55=379, P56=713, P57=174, P58=814, P59=393, P60=294
Y8 (cost = 159)	P1=160, P2=106, P4=212, P5=242, P6=106, P7=255, P8=105, P9=127, P10=84, P11=73, P12=128, P14=119, P15=326, P17=166, P18=175, P19=229, P20=272, P21=201, P22=91, P23=188, P24=142, P25=125, P26=313, P28=159, P29=209, P30=249, P31=277, P33=213, P34=162, P35=197, P37=160, P38=142, P41=140, P42=155, P44=98, P45=106, P46=230, P47=210, P49=249, P50=249, P51=256, P53=124, P56=142, P57=70, P58=285, P59=98, P60=78

Table 5.14 (Continued from Table 5.13.) PE types, their cost, and process computation time in example *random-2*. (To be continued by Table 5.15.)

PE type	process computation time
Y9 (cost = 31)	P1=956, P2=428, P3=562, P4=815, P6=435, P7=1104, P8=521, P9=941, P10=495, P12=620, P13=714, P14=634, P15=1053, P16=1010, P17=1115, P18=825, P19=903, P20=1095, P21=1195, P22=542, P23=1083, P24=562, P25=563, P26=1141, P27=897, P29=724, P31=1048, P32=1397, P33=1323, P34=855, P35=973, P36=719, P37=723, P38=730, P44=478, P45=470, P46=1445, P48=830, P50=1274, P51=1488, P53=602, P54=1204, P55=677, P56=1169, P57=330, P59=724, P60=334
Y10 (cost = 108)	P1=245, P2=167, P3=227, P4=212, P5=465, P6=175, P7=365, P8=137, P9=235, P12=218, P13=168, P14=227, P15=328, P16=286, P18=293, P19=257, P20=322, P21=425, P22=163, P23=250, P24=226, P26=289, P27=195, P29=305, P30=309, P31=301, P32=267, P33=383, P34=357, P35=354, P36=202, P37=255, P38=212, P39=157, P41=174, P42=286, P43=432, P44=150, P46=411, P47=306, P48=185, P50=265, P51=414, P52=100, P53=203, P54=287, P55=145, P58=337, P60=128
Y11 (cost = 8)	P1=2529, P2=1994, P3=2173, P4=3524, P7=4500, P8=2258, P10=1781, P11=1363, P14=2425, P15=5435, P17=4866, P18=2970, P19=5336, P20=3928, P21=3865, P22=1518, P23=3311, P24=2945, P25=2438, P26=5531, P27=3154, P29=3762, P30=5117, P31=4942, P35=5436, P36=3549, P37=2399, P38=2085, P40=3773, P41=2951, P44=1726, P45=2260, P47=3634, P49=4155, P51=4848, P52=1704, P53=2373, P54=4138, P55=2717, P56=3190, P57=1184, P58=4804, P59=2424, P60=1649
Y12 (cost = 192)	P1=118, P2=97, P4=149, P6=85, P7=188, P8=99, P9=154, P11=62, P13=123, P14=102, P15=183, P16=148, P17=225, P18=165, P19=148, P21=172, P22=64, P24=119, P25=117, P26=235, P28=106, P29=158, P31=239, P32=173, P34=192, P35=162, P36=177, P37=123, P38=136, P40=195, P41=116, P42=149, P43=260, P45=67, P46=218, P49=182, P50=170, P51=172, P53=81, P54=224, P55=71, P56=124, P57=49, P58=171, P60=64

Table 5.15 (Continued from Table 5.14.) PE types, their cost, and process computation time in example *random-2*.

6

COMMUNICATION ANALYSIS AND SYNTHESIS

6.1 OVERVIEW

Communication analysis and synthesis is an essential step in distributed system co-synthesis. Many embedded systems use custom communication topologies and the communication links are often a significant part of the system cost. This chapter describes new techniques [110] for the analysis and synthesis of the communication requirements of embedded systems during co-synthesis.

Most previous work uses two kinds of communication topologies: either a point-to-point communication link for each pair of processors, or a single bus/network for all interprocess communication. Embedded systems are application specific, and they can have ad hoc architecture with several buses, each with a different speed and a different number of PEs connected to it, according to the real-time constraints imposed on individual tasks. For instance, a bus connecting five CPUs may be implemented for light communications, while another dedicated bus is used by only two CPUs for time-critical messages. Rosebrugh and Kwang [83] described a pen-based system built from four processors of different types: a Motorola MC68331, a Motorola MC68HC05C4, a Hitachi 63484, and an Intel 8051. There are five buses with different connections for interprocessor communication in their pen-based system: power management bus, serial bus, video bus, digitizer bus, and system bus, as shown in Figure 6.1. The synthesis of communication is important for distributed embedded system design.

Communication is the bottleneck in many embedded systems, because communication links add both chip and board costs, and designers frequently underestimate peak load. Design decisions based on average communication requirements may lead to an infeasible design. The communication must be scheduled

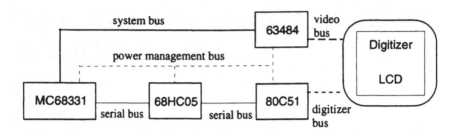

Figure 6.1 The pen-based system described by Rosebrugh and Kwang.

and allocated to determine feasibility, and communication synthesis interacts with process scheduling and engine design. According to the communication model described in Section 3.3, Section 6.2 extends the performance estimation algorithm in Chapter 4 to include communication delay. Section 6.3 then uses the delay estimates to develop methods for synthesizing communication links, based on the co-synthesis algorithm in Chapter 5. Our communication-synthesis algorithm selects the number of buses, the messages transferred on each bus, and schedules the bus communication. Section 6.4 gives the results of experiments with the algorithms.

6.2   COMMUNICATION DELAY ESTIMATION

In this section, we will propose modifications of Equation (2.2) for communication delay. The communication delay estimation is more difficult because of the interaction between PEs and communication channels; for example, Section 6.2.2 shows that preemption is no longer transitive as seen in pure PE scheduling. To simplify our description, we will ignore the techniques like phase adjustment and separation analysis mentioned in Chapter 4, though these techniques can be applied in communication analysis as well.

We first discuss how extra sending and receiving processes are added into the task graphs in the original specification to model communication in Section 6.2.1. We then derive formulas to calculate the response time for these additional communication processes in Section 6.2.2. If a CPU cannot compute and communicate at the same time, the response time of an application process in the original task graphs may be affected by the communication processes. We propose methods to deal with such a situation in Section 6.2.3.

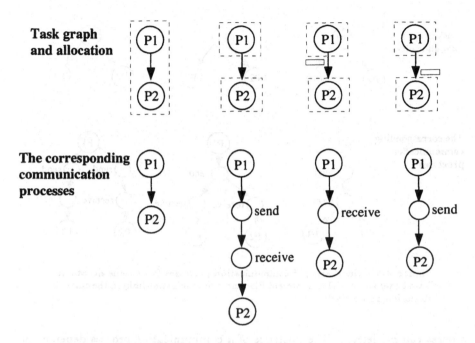

Figure 6.2 The creation of communication processes for various situations. Dash boxes represent PEs. A small solid box stands for a dual-port buffer for a PE.

Suppose the allocation and scheduling have been given. For each process P_i, c_i is the computation time or communication time, and P_i is allocated to PE_i with a priority $Pprt_i$. If P_i is a communication process, it is also allocated to BUS_i with a priority $Bprt_i$.

6.2.1 Communication Modeling

Figure 6.2 shows how a sending process and a receiving process may be inserted into an edge in a task graph for various cases. If two processes connected by an edge are allocated to the same PE, or either of the processes is a dummy process, no communication process needs to be created for this arc. When two processes connected by an edge are allocated to different PEs, at least one communication process needs to be created for the corresponding message. If there is a dual-port buffer for the process receiving the message, the receiving process is not necessary. If there is a dual-port buffer for the process sending the message, but no dual-port buffer exists for the receiving side, the sending

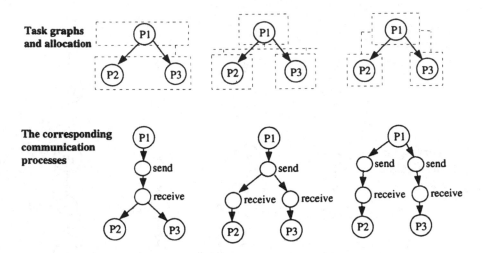

Figure 6.3 The sharing of communication processes for various situations. Dash boxes and dash line represent PEs and communication links in the current allocation, respectively.

process can be deleted. The existence of a communication process depends on the allocation of other processes.

When there is more than one edge leaving a process in a task graph, and no dual-port buffer is available, a communication process may be shared by different edges. Such cases are described in Figure 6.3. If the destinations of two edges leaving the same source process are allocated to the same PE, both the sending process and the receiving process can be shared, because once the same message is read into the local memory for the first process, the second process can also read it without accessing the bus again. If the destinations of two edges are on two different PEs which use the same bus used by the PE containing the source process, the sending process can be shared, although the receiving processes must be separated. The sharing of sending processes implied that when a process wants to broadcast data to more than one process, a bus might provide better performance than point-to-point communication links. We can reduce overhead by eliminating unnecessary communication processes,

variable	definition
d_c^C	$d_c^B + d_c^P$
	total delay of a communication process
d_c^B	b_c
	delay caused by bus activities
b_i	$x = g(x) = c_i + \sum_{P_j \in \mathcal{B}_i} c_j \cdot \lceil x/p_j \rceil$
	worst-case response time on a bus
d_c^P	$x = d_c^N + \sum_{P_j \in \mathcal{P}_c} c_j \cdot \lceil x/p_j \rceil + d_c^H(x)$
	delay caused by PE activities
d_c^N	$\max_{P_j \in \mathcal{N}_c}(b_j - 1)$
	delay caused by low-priority communication processes
$d_c^H(x)$	$\sum_{P_i \in \mathcal{H}_c} \min\{\sum_{P_j \in \mathcal{E}_i} \lceil x/p_j \rceil \cdot \lceil b_j/p_i \rceil, \lceil (x - d_c^N)/p_i \rceil\} \cdot c_i$
	delay caused by high-priority communication processes

set	definition
$\mathcal{B}_i$	$\{P_j \vert BUS_j = BUS_i, PE_i \neq PE_j, Bprt_j > Bprt_i\}$
	high-priority communication processes from other PEs
$\mathcal{P}_c$	$\{P_j \vert PE_j = PE_c, Pprt_j > Pprt_c\}$
	high-priority processes on the same PE
$\mathcal{N}_c$	$\{P_j \vert PE_j = PE_c, Pprt_j < Pprt_c, P_j \text{ is communication}\}$
	low-priority communication processes on the same PE
$\mathcal{H}_c$	$\{P_j \vert PE_j \neq PE_c, \exists P_k \in \mathcal{P}_c \text{ such that}$ $BUS_j = BUS_k \text{ and } Bprt_j > Bprt_k\}$
	high-priority communication processes from other PEs
$\mathcal{E}_i$	$\{P_j \vert P_i \in \mathcal{H}_c, P_j \in \mathcal{P}_c, BUS_j = BUS_i, Bprt_j < Bprt_i\}$
	processes in $\mathcal{P}_c$ preempted by those in $\mathcal{H}_c$

Table 6.1 The notation for the derivation of the communication process delay d_c^C.

6.2.2 Communication Process Delay

We derive the worst-case response time for a communication process in this
section. The related notation in the derivation is summarized in Table 6.1.
Example timing diagrams which demonstrate different time intervals for the
variables or sets are shown in Figure 6.4.

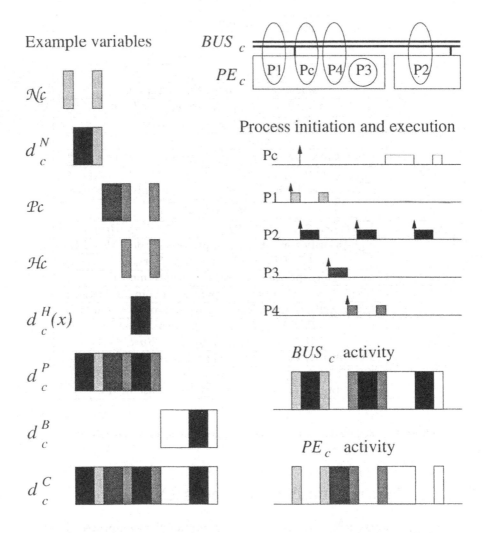

Figure 6.4 Example timing diagrams deriving the response time of P_c for the variables in Table 6.1. The priorities of processes are $P_2 > P_4 > P_c > P_1$ on BUS_c, and $P_3 > P_4 > P_c > P_1$ on PE_c.

Given any communication process P_i, define the set of processes which share the same bus as P_i, and have higher priority than P_i on the bus, but are not allocated to the same PE as P_i:

$$\mathcal{B}_i = \{P_j | BUS_j = BUS_i, PE_i \neq PE_j, Bprt_j > Bprt_i\}$$

Let the worst-case *bus response time* b_i be the longest time from the instant P_i requests the bus to the instant P_i finishes all its data transfer. Similar to Equation (2.2), b_i is the smallest positive root of the equation

$$x = g(x) = c_i + \sum_{P_j \in \mathcal{B}_i} c_j \cdot \lceil x/p_j \rceil \tag{6.1}$$

where c_j is the communication time for each communication process. The worst-case *total response time* d_c^C due to a communication process P_c is the longest time from the request of P_c to the finish of P_c. The request of a sending process occurs when the process generating the data completes its computation. The request of a receiving process occurs when the corresponding sending process finishes sending the data. The delay d_c^C is divided into two components:

$$d_c^C = d_c^P + d_c^B$$

where d_c^B is the time spent on the bus, and d_c^P is the scheduling delay to wait for some other processes to finish on the PE before starting to use the bus. Apparently, $d_c^B = b_c$ where b_c is calculated by Equation (6.1).

Define the set of processes which are allocated to the same PE as P_c, but cannot run in parallel with P_c, and have higher priority than P_c:

$$\mathcal{P}_c = \{P_j | PE_j = PE_c, Pprt_j > Pprt_c\} \tag{6.2}$$

Set $\mathcal{P}_c$ includes both application processes and communication processes when the PE spends CPU time on communication, but includes only communication processes if computation and communication can be done independently.

Based on Equation (2.2), we derive that the value of d_c^P is the smallest positive root of the equation

$$x = d_c^N + \sum_{P_j \in \mathcal{P}_c} c_j \cdot \lceil x/p_j \rceil + d_c^H(x) \tag{6.3}$$

The term d_c^N is the worst-case of the delay caused by a communication process with a lower priority than P_c on PE_c. Because a communication process is assumed to be non-preemptable, if it starts immediately (1 time unit) before the request of P_c, it will continue until it is finished even though it has lower priority. Define the set of communication processes with lower priority than P_c on the same PE:

$$\mathcal{N}_c = \{P_j | PE_j = PE_c, Pprt_j < Pprt_c, P_j \text{ is a communication process.}\}$$

If $\mathcal{N}_c = \phi$, $d_c^N = 0$. Otherwise,

$$d_c^N = \max_{P_j \in \mathcal{N}_c}(b_j - 1) \tag{6.4}$$

The function $d_c^H(x)$ in Equation (6.3) represents the worst-case total time interrupted by high-priority communication processes on buses when a communication process in P_c is running. Define the set of processes which use a bus connected to the PE for P_c and may affect the total response time of P_c:

$$\mathcal{H}_c = \{P_j | PE_j \neq PE_c, \exists P_k \in \mathcal{P}_c \text{ such that } \\ BUS_j = BUS_k \text{ and } Bprt_j > Bprt_k\}$$

For each process P_i in $\mathcal{H}_c$, define the set of processes which can be preempted by P_i on the bus, and belong to $\mathcal{P}_c$:

$$\mathcal{E}_i = \{P_j | P_i \in \mathcal{H}_c, P_j \in \mathcal{P}_c, BUS_j = BUS_i, Bprt_j < Bprt_i\}$$

The function $d_c^H(x)$ in Equation (6.3) is formulated as follows:

$$d_c^H(x) = \sum_{P_i \in \mathcal{H}_c} \min\{\sum_{P_j \in \mathcal{E}_i} \lceil x/p_j \rceil \cdot \lceil b_j/p_i \rceil, \lceil (x - d_c^N)/p_i \rceil\} \cdot c_i \tag{6.5}$$

The formulation is special because the preemptive relationship is *not transitive* when two resources—PE and bus—are involved. In fixed-priority scheduling of a single CPU, if process P_1 can preempt P_2, and P_2 can preempt P_3, P_1 is allowed to preempt P_3 too. Suppose P_1 and P_2 are communication processes and use the same bus but different PEs, while P_2 and P_3 use the same CPU but P_3 does not use buses. If P_1 can preempt P_2 during bus transactions, and P_2 can preempt P_3 on the CPU, it is still impossible for P_1 to preempt P_3 because they use different resources.

For a process $P_i \in \mathcal{H}_c$, it can preempt a process P_j in $\mathcal{E}_i$ at most $\lceil b_j/p_i \rceil$ times, where b_j is the longest time P_j stays on the bus and can be solved from Equation (6.1). Process P_j can occur during the time interval d_c^P at most $\lceil x/p_j \rceil$ times. Therefore, P_i can affect d_c^P at most $\lceil x/p_j \rceil \cdot \lceil b_j/p_i \rceil$ times, as demonstrated by Figure 6.5. On the other hand, in addition to the occurrences during the interval d_c^N, the number of occurrences for P_i cannot exceed $\lceil (x - d_c^N)/p_i \rceil$, as seen in Figure 6.6. Consequently, we use a *min* function for these two formulas. Note that a *min* function is a non-decreasing function, so fixed-point iterations will still converge for Equation (6.3).

6.2.3 Application Process Delay

If the computation and communication cannot be executed in parallel on a PE, the communication processes can cause extra delay in the response time of an application process when the delay estimation algorithm visits it. The worst-case response time d_a^A for an application process P_a is the smallest positive root of the equation

$$x = g(x) = d_a^N + c_a + \sum_{P_j \in \mathcal{P}_a} c_j \cdot \lceil x/p_j \rceil + d_a^H(x) \qquad (6.6)$$

The definition of set $\mathcal{P}_a$ is the same as that of $\mathcal{P}_c$ in (6.2). Equation (6.6) is similar to Equation (6.3), but the computation time c_a is included. The calculation of d_a^N is similar to that of d_c^N in Equation (6.4), and the formulation of $d_a^H(x)$ is similar to that in Equation (6.5).

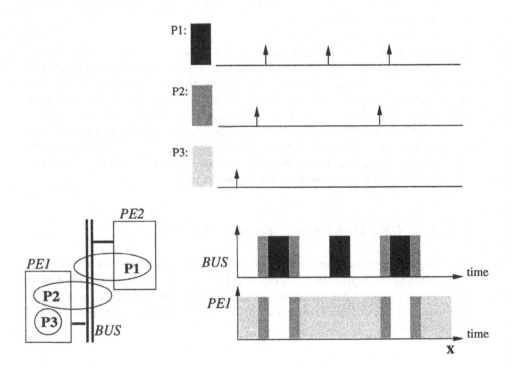

Figure 6.5 The execution of processes on a bus connected with two PEs. Suppose P_2 has higher priority than P_3 on PE_1, and P_1 has higher priority than P_2 on the bus. Let x be the total response time of P_3. In this case (the communication time of P_2 is shorter), the number of times P_1 affects the response time of P_3 is $\lceil x/p_2 \rceil \cdot \lceil b_2/p_1 \rceil$.

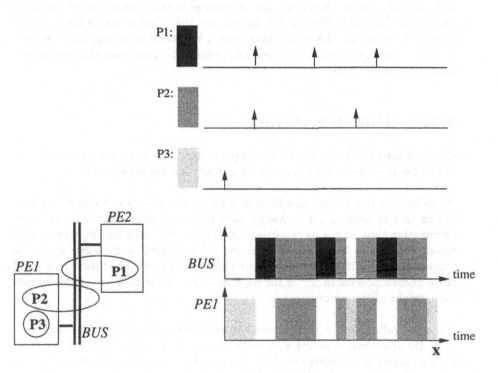

Figure 6.6   The execution of processes on a bus connected with a PE in two cases. Suppose P_2 has higher priority than P_3 on a PE, and P_1 has higher priority than P_2 on a bus. Let x be the total response time of P_3. In this case (the communication time of P_2 is longer), the number of times P_1 affects the response time of P_3 is $\lceil x/p_1 \rceil$.

6.3   COMMUNICATION SYNTHESIS

Our communication co-synthesis algorithm uses an iterative improvement strategy like the algorithm in Chapter 5. At each step, in addition to the reallocation of a process or the creation of a new PE, the algorithm may also reallocate one message from one bus to another or create a new bus. System performance and cost will change when a process or message is reallocated. Using the delay estimation method in Section 6.2, we can get the value of a task delay in the current solution and the delay value after a reallocation to evaluate the reallocation. Sensitivity analysis is used to evaluate the candidate changes for the candidate architecture.

6.3.1   Bus Scheduling

Based on the priority prediction technique in Section 5.3, we develop a heuristic to use the inverse-deadline priority assignment for bus scheduling.

We define the **fractional deadline** of a communication process in a way similar to that in Section 5.3.1. Assign each process a weight equal to its latest termination time minus its latest initiation time. For each bus B, temporarily assign weight zero to the set of processes $\mathcal{J}^B$ allocated to B. Then apply the longest-path algorithm backwards from the end of a task. The **latest required time** of each process in $\mathcal{J}^B$ is the hard deadline of the task minus the longest path weight of the process.

If the bus arbitration scheme allows the assignment of a priority for each message, the fractional deadline d_i of each process $P_i \in \mathcal{J}^B$ is its latest required time minus its latest initiation time. We can then order the priority by d_i—the shorter the fractional deadline, the higher its priority. When the bus arbitration scheme assigns priorities to PEs only, and all the communication processes on the same PE have the same priority on the bus, the PEs on a bus are scheduled as follows. For each PE R on the bus B and for each task τ, define $\mathcal{J}_R^B$ as the subset of $\mathcal{J}^B$ such that $\mathcal{J}_R^B$ contains only those processes which are allocated to R. Let the largest of the latest required times for processes in $\mathcal{J}_R^B$ be the fractional deadline of all the communication processes in task τ for $\mathcal{J}_R^B$. Choose the tightest (smallest) fractional deadline among the fractional deadlines computed in all the task graphs as the deadline of the PE. Schedule the PEs on the bus by assigning a higher priority to a PE with smaller deadline.

6.3.2 Communication Synthesis

We deal with the communication by the following modifications to the algorithm in Section 5.5:

- In the initial solution, implement a bus for each message. Because we allocate a different PE for each process in the initial solution, there is a bus for each edge in a task graph.

- When computing sensitivities, include communication delays computed by the method in Section 6.2, and the bus cost mentioned in Section 3.3. Apply the idle-PE elimination technique in Section 5.4.1 for the least-utilized bus in the cost calculation.

- In addition to considering possible reallocation of a process to another PE, consider also possible reallocation of a message to another bus. Choose either a process reallocation or a communication reallocation according to sensitivity analysis during each iteration. Exclude the choices whose sensitivities are negative.

- When no reallocation remains feasible, try to create a bus, in addition to a new PE. Compute the sensitivities for each possible bus type and each message.

- The reallocation with highest sensitivity will be chosen. Make such reallocation, delete and regenerate communication processes for the new allocation, according to the approach in Section 6.2.1. Reschedule the PEs and buses.

- Repeat the above steps until no reallocation is possible.

6.4 EXPERIMENTAL RESULTS

We implemented our algorithm in C++ and performed experiments on several examples. All experiments were performed on a Sun Sparcstation SS20. Some results are summarized in Table 6.3.

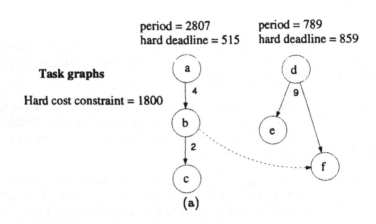

Task graphs

Hard cost constraint = 1800

period = 2807   period = 789
hard deadline = 515   hard deadline = 859

(a)

PE type	Cost	Computation time					
		a	b	c	d	e	f
X	$800	179	95	100	213	367	75
Y	$500	204	124	173	372	394	84
Z	$400	210	130	193	399	494	91

(b)

Bus type	communication time per data unit	Bus interface cost		
		X	Y	Z
B1	2	36	19	30
B2	1	20	10	15

Figure 6.7 A small example *ex2*. (a) The task graphs for two tasks, their periods and deadlines. The number below a process is the volume of data for transfer. The dash line from *b* to *f* is an inter-task communication. (b) The computation time of each process and the cost on each type of PE. (c) The communication time per data unit and the bus interface cost on each type of bus.

Step	embedded system architecture	cost
1	(Y: e) (Z: a) (Z: b) (Z: c) (Z: d) (Z: f) (B1: a→b) (B1: b→c) (B1: d→e) (B1: d→f) (B1: b→f)	$2645
2	(Y: e d) (Z: a) (Z: b) (Z: c) (Z: f) (B1: a→b) (B1: b→c) (B1: d→f) (B1: b→f)	$2215
3	(Y: e d) (Z: b a) (Z: c) (Z: f) (B1: b→c) (B1: d→f) (B1: b→f)	$1785
4	(Y: e d) (Z: c b a) (Z: f) (B1: d→f) (B1: b→f)	$1355
5	(X: c) (Y: e d) (Z: b a) (Z: f) (B1: b→c) (B1: d→f) (B1: b→f)	$2190
6	(X: c f) (Y: e d) (Z: b a) (B1: b→c b→f) (B1: d→f)	$1765
7	(X: c f b) (Y: e d) (Z: a) (B1: a→b) (B1: d→f)	$1765

Table 6.2 The iterative optimization for *ex2*. After each step, each PE or bus in the refined system architecture is shown by its type, followed by a colon, and then the processes or messages allocated to it are listed from the highest priority to the lowest according to the schedule.

6.4.1 A Step-by-Step Example

The first example, *ex2* is a small example shown in Figure 6.7. The soft cost constraints and soft deadlines are set to zero to encourage optimization whenever it is possible. The communication time is assumed to be proportional to the size of data. Communication and computation cannot run in parallel on all types of PEs. The embedded system architecture, total cost and the satisfaction of real-time deadlines after each iterative step are given in Table 6.2.

6.4.2 Prakash and Parker's Examples

Our second example is the second example of Prakash and Parker [77]; their algorithm works on a single task graph. We use their assumption on bus-style interconnection; the cost of the system is dominated by the cost of the processors selected; each PE can perform computation and communication in parallel. We compare our results with theirs in Table 6.4 for three designs with different hard cost constraints. Our result for the task delay is close to their

Example	The result			CPU
	#PE	#bus	cost	time
ex1	3	2	$1765	10.63s
prakash-parker-3	see Table 6.4			
prakash-parker-4	3	1	$11.5	193.3s

Table 6.3 The final result and the CPU time of running our algorithm for each example.

Design	Hard cost constraint	Prakash & Parker's		Our results	
		delay	CPU time	delay	CPU time
1	$10	6	107.30min	6	57.3sec
2	$6	7	89.53min	8	57.58sec
3	$5	15	61.52min	-	56.70sec

Table 6.4 The results for Prakash and Parker's bus examples. Prakash and Parker's run times were measured on a Solbourne 5e/900 (similar to Sun SPARC 4/490). Our run times were measured on a Sun Sparc20 Workstation. For Design 3, our algorithm cannot give a solution satisfying the hard cost constraint.

result, but the efficiency of our algorithm is better. Their algorithm is unable to handle more complex communication models, such as the cases where the cost of buses are significant, computation and communication have resource conflict, or there are multiple buses.

We also combine Prakash and Parker's two examples together and assign the periods as well as the deadlines of 7 and 15 to the two tasks. The results are given under *prakash-parker-4*. This example demonstrates how our algorithm can co-synthesize from multiple disjoint task graphs for communication.

6.4.3   Random Large Examples

In order to test the performance of our algorithm on larger examples, we randomly generate a example *random-3*, whose results are shown in Table 6.5 and Table 6.6. The system specification of *random-3* is given in Table 6.7, Table 6.8,

Example	Problem size			The result			CPU
	#task	#process	#PE type	#PE	#BUS	cost	time
random-3	4	30	7	14	8	$1628	23779s

Table 6.5 The problem size, the final result, and the CPU time of running the communication synthesis algorithm for a randomly generated example.

Task	worst-case delay
τ_1	1153
τ_2	1058
τ_3	712
τ_4	1141

Table 6.6 Task worst-case delays in the result of *random-3*.

Task	period	soft deadline	hard deadline
τ_1	340986	0	6433
τ_2	318366	0	6006
τ_3	493914	0	9319
τ_4	395132	0	7455

Hard cost constraint = 5880
Soft cost constraint = 53
Bus cost = 11

Table 6.7 Task periods and deadlines in example *random-3*.

Table 6.9, and Table 6.10. It took a few hours for our algorithm to handle this example. Compared with the results in Section 5.6.4, communication synthesis takes much more time. In addition to the processes specified by the designer, there are many other extra communication processes, and the actual size of the problem becomes much larger. For example, *random-3* has 30 processes, but the communication synthesis algorithm needs to schedule and allocate more than 120 processes. The performance analysis algorithm is more time-consuming for communication due to the complex interaction between PEs and buses.

Task	data dependency edges
τ_1	P6/P7, P2/P5, P3/P6, P1/P5, P3/P7, P4/P6, P1/P2, P1/P7, P1/P6, P2/P3, P4/P7, P4/P6, P3/P5, START/P1, START/P4, P5/END, P7/END
τ_2	P12/P14, P8/P11, P11/P12, P11/P14, P9/P11, P12/P13, P10/P13, P8/P14, P11/P13, P9/P14, START/P8, START/P9, START/P10, P13/END, P14/END
τ_3	P15/P17, P19/P21, P18/P21, P15/P22, P19/P20, P16/P18, P20/P21, P16/P17, START/P15, START/P16, P17/END, START/P19, P21/END, P22/END
τ_4	P24/P26, P24/P28, P24/P27, P28/P29, P27/P29, P26/P27, P25/P26, P27/P28, P23/P24, P23/P28, P23/P25, P24/P29, P29/P30, P25/P29, START/P23, P30/END

Table 6.8 Task graphs in example *random-3*.

PE type	process computation time
Y1 (cost = 126)	P3=180/2, P4=172/2, P5=359/3, P6=125/2, P7=251/3, P8=121/1, P9=204/3, P10=137/1, P11=86/1, P12=221/3, P13=186/2, P15=345/3, P18=298/3, P19=316/1, P21=290/2, P25=194/1, P26=343/2, P27=243/1, P28=185/1, P29=193/2, P30=242/1
Y2 (cost = 107)	P1=286/2, P2=156/1, P3=226/1, P4=242/3, P5=408/3, P6=131/2, P7=271/1, P9=212/2, P10=119/1, P11=104/1, P12=170/2, P13=154/2, P14=181/1, P16=300/1, P17=349/2, P18=231/2, P19=248/2, P20=389/4, P21=355/3, P23=322/1, P24=184/2, P25=247/3, P26=456/5, P27=209/2, P28=165/2
Y3 (cost = 196)	P1=142/1, P3=98/1, P6=77/1, P7=172/1, P8=101/1, P9=151/2, P11=58/1, P12=98/1, P13=115/1, P14=118/2, P15=193/2, P17=152/2, P18=184/1, P19=219/1, P20=150/1, P21=152/2, P22=92/1, P23=118/1, P24=110/1, P25=96/1, P26=246/3, P28=140/2, P29=161/2, P30=141/2
Y4 (cost = 104)	P2=138/2, P3=210/2, P4=249/1, P5=407/3, P6=154/1, P7=321/2, P9=199/1, P10=157/2, P11=116/2, P13=211/2, P14=220/3, P15=374/2, P16=326/2, P17=391/1, P19=414/2, P20=364/2, P21=395/3, P22=168/2, P23=260/3, P26=321/4, P28=213/1, P29=271/2, P30=273/1

Table 6.9 PE types, their cost, process computation time, and message communication time in example *random-3*. P3=180/2, for example, represents that the computation time of process P3 is 180 and the message communication time out of P3 is 2. The computation and communication can be performed in parallel. (To be continued by Table 6.10.)

PE type	process computation time
Y5 (cost = 53)	P1=384/2, P2=335/4, P3=437/1, P4=630/3, P6=228/1, P7=713/1, P9=456/2, P11=181/1, P12=493/1, P13=305/3, P14=313/2, P15=612/6, P16=604/4, P17=777/6, P18=674/4, P21=744/5, P23=691/3, P24=464/5, P25=479/4, P26=607/4, P27=420/3, P29=435/2
Y6 (cost = 69)	P1=427/4, P2=202/3, P3=326/1, P4=411/5, P5=693/4, P6=186/2, P7=389/3, P8=306/4, P9=440/1, P11=203/3, P12=379/4, P14=250/3, P15=649/4, P16=323/4, P17=611/6, P18=432/3, P19=590/6, P21=571/1, P22=257/2, P23=523/5, P24=329/1, P25=244/1, P26=509/6, P27=306/1, P28=388/2, P29=373/3, P30=577/6
Y7 (cost = 156)	P1=131/2, P2=101/1, P3=145/1, P4=139/2, P5=315/4, P6=91/1, P7=211/2, P9=147/1, P10=99/1, P12=132/2, P13=140/2, P14=152/1, P15=325/2, P16=210/3, P18=163/2, P20=266/1, P21=213/2, P22=98/1, P23=233/1, P25=104/1, P26=258/3, P28=179/2, P29=164/2, P30=245/3

Table 6.10   (Continued from Table 6.9.) PE types, their cost, process computation time, and message communication time in example *random-3*.

7

CONCLUSIONS

7.1 CONTRIBUTIONS

Distributed computers are often the most cost-effective means of meeting the performance requirements of an embedded computing application. Embedded distributed computing is a particularly challenging design problem because the hardware and software architectures must be simultaneously optimized. We have presented a suit of new co-synthesis algorithms for heterogeneous distributed systems of arbitrary topology. Our algorithms make use of tight yet easy-to-compute bounds on the schedulability of tasks to guide its heuristic search.

The major contribution of Chapter 4 is an analytic performance estimation algorithm for multiple-rate distributed systems—there are few analytic algorithms which do not use LCM enumeration, and our algorithm is much more accurate than any other previous analytic algorithm. Specifically, Chapter 4 presents new techniques which

- apply fixed-point iterations to replace rate-monotonic analysis;

- use phase adjustment to reduce duplicate preemptions;

- use separation analysis to avoid false preemptions.

The major contribution of Chapter 5 is a sensitivity-driven co-synthesis algorithm—very few co-synthesis algorithms have been able to deal with distributed systems. Our co-synthesis algorithm offers better ability than other distributed synthesis algorithms in the following ways:

- The choice of optimization is guided by sensitivity analysis instead of first-come-first-served basis.

- It performs preemptive priority scheduling for multiple-rate systems during co-synthesis.

- Two-stage optimization is used to eliminate poor-utilized PEs.

The major contribution of Chapter 6 is the first algorithm to analyze and synthesize custom communication topology for embedded systems—regular architectures such as point-to-point communication and single-bus connection are not suitable for embedded systems. The algorithm is able to

- handle the complex interaction between the two resources: PEs and communication channels;

- simultaneously allocate and schedule communication channels.

We believe that algorithms such as those in this book will be eventually incorporated as an important tool for the practicing embedded system designers.

7.2   FUTURE DIRECTIONS

In order to develop a complete embedded system co-synthesis tool for real industry designs, there are several areas where work needs to be done.

7.2.1   Specification Languages

Everyone has his or her preference on either *text-based* or *graph-based* formats of specification. Graph-based specification is frequently used as intermediate models for CAD tools to operate on and as output formats to display the system. However, text-based specification is more prevalent for human input: most large software systems are specified by programming languages like C rather than flow charts; VHDL and Verilog are now more popular than schematic entry for complex hardware systems.

VHDL and Verilog are originally simulation languages, so the complete language set is not all synthesizable. Current VHDL or Verilog synthesis tools

Since all the sensitivity values are negative, no movement can be adopted, and the current system configuration becomes the final embedded system architecture, where the hard cost constraint is violated and no feasible design is found.

On the other hand, if we use formula (5.2) to modify the cost calculation for idle-PE elimination, the sensitivities are

- Move P_1 from R_1 to R_2:

    $$\mathbf{v} = (10, 10, 20, 20, 10.8)$$

    $$\mathbf{D} = (0, -0.196, 0.192, 0.189, -0.72)$$

    $$S = 0.618409$$

- Move P_2 from R_1 to R_2:

    $$\mathbf{v} = (10, 10, 20, 20, 10.82)$$

    $$\mathbf{D} = (0, -0.196, 0.192, 0.189, -0.706)$$

    $$S = 0.605827$$

- Move P_3 from R_2 to R_1:

    $$\mathbf{v} = (10, 20, 30, 10, 10.85)$$

    $$\mathbf{D} = (0, 0, 0.385, -0.189, -0.692)$$

    $$S = 0.614913$$

- Move P_4 from R_2 to R_1:

    $$\mathbf{v} = (10, 20, 10, 30, 10.87)$$

    $$\mathbf{D} = (0, 0, 0, 0.187, -0.679)$$

    $$S = 0.552449$$

slow and does not guarantee better results. However, based on our current approach, we believe that there can be a lot of enhancements in the future. In particular, the following aspects may be considered:

- A good initial solution may lead to faster convergence and a better final design. The initial solution we use now produces too many PEs at the beginning and slows down the convergence time.

- A more optimal scheduling algorithm for multiple-rate distributed system may improve the final system performance a lot.

- More heuristics are necessary to prevent the solution from stopping at local minima too often.

In addition, given a specification in a high-level language, a *retargetable* compiler is important to map the specification to various types of PEs, either CPUs or ASICs.

7.2.4   Communication

Communication is critical for the performance and cost of distributed systems. The algorithm in Chapter 6 is based on the performance estimation algorithm and the co-synthesis algorithm, so it can be improved in a similar way. In addition, we think the following improvements are important for communication synthesis in the future:

- Cache effects can affect the interaction between CPU and communication channels. Li, Malik and Wolfe's algorithm [59] and others which analyze caching effects can be used to adjust available bus bandwidth.

- A more accurate model for more communication protocols is crucial. For instance, a serial bus controlled by interrupt service routines produce several small periodic processes for a single message, where each communication process transfers one byte of data. How to intelligently partition a long message into small chunks is also an interesting problem.

- Many modern microprocessor has on-chip dual-port memory, on-chip cache, more than one address bus and data bus, etc. The size of on-chip dual-port memory is limited. If the communication data are fewer than the dual-port memory size, the microprocessor may perform computation and communication in parallel; but if the communication data are too large, computation will be suspended by communication. As a result, it is difficult to conclude whether a CPU type can perform computation and communication in parallel or not; it depends on the volume of communication data. When the volume of communication data is large, how to utilize the on-chip memory efficiently is important. Similarly, if a CPU has two address/data bus but the communication channels it needs to access are more than two, how to assign the communication channels to the address/data buses affects both system performance and cost.

REFERENCES

[1] Alauddin Alomary, Takeharu Nakata, Yoshimichi Honma, Jun Sato, Nobuyuki Hikichi, and Masaharu Imai. PEAS-I: A hardware/software co-design system for ASIPs. In *Proceedings of European Design Automation Conference*, 1993.

[2] Tod Amon, Henrik Hulgaard, Steven M. Burns, and Gaetano Borriello. An algorithm for exact bounds on the time separation of events in concurrent systems. In *Proceedings of IEEE International Conference on Computer Design*, pages 166–173, 1993.

[3] S. Antoniazzi, A. Balboni, W. Fornaciari, and D. Sciuto. Hardware/software codesign for embedded telecom systems. In *Proceedings of IEEE International Conference on Computer Design*, pages 278–281, 1994.

[4] S. Antoniazzi, A. Balboni, W. Fornaciari, and D. Sciuto. A methodology for control-dominated systems codesign. In *Proceedings of International Workshop on Hardware-Software Co-Design*, pages 2–9, 1994.

[5] Apple Computer, Inc. *Designing Cards and Drivers for the Macintosh Family*. Addison-Wesley Publishing Company, 1992.

[6] Guido Araujo and Sharad Malik. Optimal code generation for embedded memory non-homogeneous register architectures. In *Proceedings of 8th International Symposium on System Synthesis*, pages 36–41, 1995.

[7] M. Auguin, M. Belhadj, J. Benzakki, C. Carriere, G. Durrieu, T. Gautier, M. Israel, P. Le Guernic, M. Lemaitre, E. Martin, P. Quinton, L. Reideau, F. Rousseau, and P. Sentieys. Towards a multi-formalism framework for architectural synthesis: The ASAR project. In *Proceedings of International Workshop on Hardware-Software Co-Design*, pages 2–9, 1994.

[8] E. Barros, W. Rosenstiel, and X. Xiong. A method for partitioning UNITY Language in hardware and software. In *Proceedings of European Design Automation Conference*, pages 220–225, 1994.

[9] G. Borriello and R. Katz. Synthesis and optimization of interface transducer logic. In *Proceedings of IEEE International Conference on Computer-Aided Design*, 1987.

[10] Gaetano Borriello. *A new interface specification methodology and its application to transducer synthesis*. PhD thesis, University of California, Berkeley, May 1988.

[11] A. W. Both, B. Biermann, R. Lerch, Y. Manoli, and K. Sievert. Hardware-software codesign of application specific microcontrollers. In *Proceedings of European Design Automation Conference*, pages 72–77, 1994.

[12] J. A. Brzozowski, T. Gahlinger, and F. Mavaddat. Consistency and satisfiability of waveform timing specification. *Networks*, pages 91–107, January 1991.

[13] Klaus Buchenrieder and Christian Veith. A prototyping environment for control-oriented HW/SW systems using state-charts, activity-charts and FPGA's. In *Proceedings of European Design Automation Conference*, pages 60–65, 1994.

[14] J. Buck, S. Ha, , E. A. Lee, and D. G. Messerschmitt. Ptolemy: A framework for simulating and prototyping heterogeneous systems. *International Journal of Computer Simulation*, January 1994.

[15] Richard L. Burden and J. Douglas Faires. *Numerical Analysis*. PWS publishers, third edition, 1985.

[16] Timothy Burks and Karem Sakallah. Min-max linear programming and the timing analysis of digital circuits. In *Proceedings of IEEE International Conference on Computer-Aided Design*, pages 152–155, 1993.

[17] Raul Camposano. Path-based scheduling for synthesis. *IEEE Transactions on Computer-Aided Design of Integrated Circuits and Systems*, 10(1):85–93, January 1991.

[18] M. Chiodo, P. Guisto, H. Hsieh, A. Jurecska, L. Lavagno, and A. Sangiovanni-Vincentelli. A formal specification model for hardware-software codesign. In *Proceedings of International Workshop on Hardware-Software Co-Design*, 1993.

[19] M. Chiodo, P. Guisto, H. Hsieh, A. Jurecska, L. Lavagno, and A. Sangiovanni-Vincentelli. Synthesis of mixed software-hardware implementations from cfsm specifications. In *Proceedings of International Workshop on Hardware-Software Co-Design*, 1993.

[20] Massimiano Chiodo, Paolo Guisto, Attila Jurecska, Harry C. Hsieh, Alberto Sangiovanni-Vincentelli, and Luciano Lavagno. Hardware-software codesign of embedded systems. *IEEE MICRO*, 14(4):26–36, August 1994.

[21] Pai Chou, Elizabeth A. Walkup, and Gaetano Borriello. Scheduling for reactive real-time systems. *IEEE MICRO*, 14(4):37–47, August 1994.

[22] Wesley W. Chu and Lance M-T. Lan. Task allocation and precedence relations for distributed real-time systems. *IEEE Transactions on Computers*, C-36(6):667–679, June 1987.

[23] Wesley W. Chu, Chi-Man Sit, and Kin K. Leung. Task response time for real-time distributed systems with resource contentions. *IEEE Transactions on Software Engineering*, 17(10):1076–1092, October 1991.

[24] Thomas H. Cormen, Charles E. Leiserson, and Ronald L. Rivest. *Introduction to Algorithms*. McGraw-Hill, 1990.

[25] Sari L. Coumeri and Donald E. Thomas. A simulation environment for hardware-software codesign. In *Proceedings of IEEE International Conference on Computer Design*, pages 58–63, 1995.

[26] Joseph G. D'Ambrosio and Xiaobo Hu. Configuration-level hardware/software partitioning for real-time embedded systems. In *Proceedings of International Workshop on Hardware-Software Co-Design*, pages 34–41, 1994.

[27] Rolf Ernst, Jörg Henkel, and Thomas Benner. Hardware-software cosynthesis for microcontrollers. *IEEE Design & Test of Computers*, 10(4), December 1993.

[28] Tony Gahlinger. *Coherence and satisfiability of waveform timing specification*. PhD thesis, University of Waterloo, May 1990.

[29] Richard Gerber, William Pugh, and Manas Saksena. Parametric dispatching of hard real-time tasks. *IEEE Transactions on Computers*, 44(3):471–479, March 1995.

[30] Rajesh K. Gupta. *Co-synthesis of Hardware and Software for Digital Embedded Systems*. PhD thesis, Stanford University, December 1993.

[31] Rajesh K. Gupta and Giovanni De Micheli. Hardware-software cosynthesis for digital systems. *IEEE Design & Test of Computers*, 10(3):29–41, September 1993.

[32] Rajesh K. Gupta and Giovanni De Micheli. Constrained software generation for hardware-software systems. In *Proceedings of International Workshop on Hardware-Software Co-Design*, pages 56–63, 1994.

[33] Manjote S. Haworth, William P. Birmingham, and Daniel E. Haworth. Optimal part selection. *IEEE Transactions on Computer-Aided Design of Integrated Circuits and Systems*, CAD-12(10):1611–1617, October 1993.

[34] Jörg Henkel, Rolf Ernst, Ullrich Holtmann, and Thomas Benner. Adaption of partitioning and high-level synthesis in hardware/software co-synthesis. In *Proceedings of IEEE International Conference on Computer-Aided Design*, pages 96–100, 1994.

[35] D. Herrmann, Jörg Henkel, and Rolf Ernst. An approach to the adaption of estimated cost parameters in the COSYMA system. In *Proceedings of International Workshop on Hardware-Software Co-Design*, pages 100–107, 1994.

[36] C. J. Hou and K. G. Shin. Allocation of periodic task modules with precedence and deadline constraints in distributed real-time systems. In *Proceedings of Real-Time Systems Symposium*, pages 157–169, 1982.

[37] Xiaobo (Sharon) Hu, Joseph G. D'Ambrosio, Brian T. Murray, and Dah-Lain (Almon) Tang. Codesign of architectures for automotive powertrain modules. *IEEE MICRO*, 14(4):17–25, August 1994.

[38] Ing-Jer Huang and Alvin M. Despain. Synthesis of application specific instruction sets. *IEEE Transactions on Computer-Aided Design of Integrated Circuits and Systems*, 14(6), June 1995.

[39] C.-T. Hwang, J.-H. Lee, and Y.-C. Hsu. A formal approach to the scheduling problem in high-level synthesis. *IEEE Transactions on Computer-Aided Design of Integrated Circuits and Systems*, 10(4):464–475, April 1991.

[40] T. Ismail, K. O'Brien, and A. Jerraya. Interactive system-level partitioning with PARTIF. In *Proceedings of the European Conference on Design Automation*, 1994.

[41] T. B. Ismail, M. Abid, and A. Jerraya. COSMOS: A codesign approach for communicating systems. In *Proceedings of International Workshop on Hardware-Software Co-Design*, pages 17–24, 1994.

[42] Axel Jantsch, Peeter Ellervee, Johnny Öberg, Ahmed Hermani, and Hannu Tenhumen. Hardware/software partition and minimizing memory

interface traffic. In *Proceedings of European Design Automation Conference*, pages 226–231, 1994.

[43] Asawaree Kalavade and Edward A. Lee. A hardware-software codesign methodology for DSP applications. *IEEE Design & Test of Computers*, 10(3):16–28, September 1993.

[44] Asawaree Kalavade and Edward A. Lee. A global criticality/local phase driven algorithm for the constrained hardware/software partition problem. In *Proceedings of International Workshop on Hardware-Software Co-Design*, pages 42–48, 1994.

[45] Asawaree P. Kalavade. *System-Level Codesign of Mixed Hardware-Software Systems*. PhD thesis, University of California, Berkeley, September 1995.

[46] Brian W. Kernighan and S. Lin. An efficient heuristic procedure for partitioning graphs. *Bell System Technical Journal*, 49(2):291–308, 1970.

[47] Kurt Keutzer. Hardware-software co-design and ESDA. In *Proceedings of Design Automation Conference*, pages 435–436, 1994.

[48] G. Koch, U. Kebschull, and W. Rosenstiel. A prototyping environment for hardware/software codesign in the COBRA project. In *Proceedings of International Workshop on Hardware-Software Co-Design*, pages 10–16, 1994.

[49] David Ku and Giovanni De Micheli. Relative scheduling under timing constraints. *IEEE Transactions on Computer-Aided Design of Integrated Circuits and Systems*, 11(6):696–717, June 1992.

[50] Jainendra Kumar, Noel Strader, Jeff Freeman, and Michael Miller. Emulation verification of the Motorola 68060. In *Proceedings of IEEE International Conference on Computer Design*, pages 150–158, 1995.

[51] L. Lavagno and A. Sangiovanni-Vincentelli. Linear programming for optimum hazard elimination in asynchronous circuits. In *Proceedings of IEEE International Conference on Computer Design*, 1992.

[52] E. A. Lee and D. G. Messerschmitt. Static scheduling of synchronous data flow programs for digital signal processing. *IEEE Transactions on Computers*, 36(1):24–35, January 1987.

[53] John Lehoczky, Lui Sha, and Ye Ding. The rate monotonic scheduling algorithm: Exact characterization and average case behavior. In *Proceedings of Real-Time Systems Symposium*, pages 166–171, 1989.

[54] John P. Lehoczky and Lui Sha. Performance of real-time bus scheduling algorithms. *ACM Performance Evaluation Review*, May 1986.

[55] Dennis W. Leinbaugh and Mohammed-Reza Yamani. Guaranteed response times in a distributed hard-real-time environment. In *Proceedings of Real-Time Systems Symposium*, pages 157–169, 1982.

[56] Joseph Y.-T. Leung and Jennifer Whitehead. On the complexity of fixed-priority scheduling of periodic, real-time tasks. *Performance Evaluation*, 2:237–250, 1982.

[57] Steven Li and Sharad Malik. Performance analysis of embedded software using implicit path enumeration. In *Proceedings of Design Automation Conference*, 1995.

[58] Steven Li and Ping-Wen Ong. Cinderella for DSP3210. Unpublished report., 1994.

[59] Yau-Tsun Steven Li, Sharad Malik, and Andrew Wolfe. Performance estimation of embedded software with instruction cache modeling. In *Proceedings of IEEE International Conference on Computer-Aided Design*, pages 380–387, 1995.

[60] Y.-Z. Liao and C. K. Wong. An algorithm to compact a VLSI symbolic layout with mixed constraints. *IEEE Transactions on Computer-Aided Design of Integrated Circuits and Systems*, CAD-2(2):62–69, April 1983.

[61] C. L. Liu and James W. Layland. Scheduling algorithms for multiprogramming in a hard-real-time environment. *Journal of the Association for Computing Machinery*, 20(1):46–61, January 1973.

[62] Leo Yuhsiang Liu and R. K. Shyamasundar. Static analysis of real-time distributed systems. *IEEE Transactions on Software Engineering*, 16(4):373–388, April 1990.

[63] James D. Lyle. *SBus: Information, Applications, and Experience*. Springer-Verlag, 1992.

[64] L. Maliniak. Logic emulator meets the demands of CPU designers. *Electronic Design*, April 1993.

[65] Michael C. McFarland, Alice C. Parker, and Raul Camposano. The high-level synthesis of digital systems. *Proceedings of the IEEE*, 78(2):301–318, February 1990.

[66] Kenneth McMillan and David Dill. Algorithms for interface timing veri-fication. In *Proceedings of IEEE International Conference on Computer Design*, pages 48–51, 1992.

[67] G. D. Micheli, D. C. Ku, F. Mailhot, and T. Truong. The olympus synthesis system for digital design. *IEEE Design and Test Magazine*, pages 37–53, October 1990.

[68] Giovanni De Micheli. Guest editor's introduction: Hardware-software codesign. *IEEE MICRO*, 14(4):8–9, August 1994.

[69] Aloysius K. Mok, Prasanna Amerasinghe, Moyer Chen, and Kamtron Tantisirivat. Evaluating tight execution time bounds of programs by annotations. In *Proceedings of IEEE Workshop on Real-Time Systems Operating Systems and Software*, pages 74–80, 1989.

[70] Chris Myers and Teresa H. Y. Meng. Synthesis of timed asynchronous circuits. In *Proceedings of IEEE International Conference on Computer Design*, pages 279–284, 1992.

[71] Kunle A. Olukotun, Rachid Helaibel, Jeremy Levitt, and Ricardo Ramirez. A software-hardware cosynthesis approach to digital system simulation. *IEEE MICRO*, 14(4):48–58, August 1994.

[72] Christos H. Papadimitriou and Kenneth Steiglitz. *Combinatorial opti-mization: algorithms and complexity*. Prentice-Hall, 1982.

[73] Chang Yun Park. *Predicting Deterministic Execution Times of Real-Time Programs*. PhD thesis, University of Washington, Seattle, August 1992.

[74] David A. Patterson and John L. Hennessy. *Computer Organization and Design: The Hardware/Software Interface*. Morgan Kaufmann Publishers Inc., 1994.

[75] P. G. Paulin and J. P. Knight. Force-directed scheduling for the behav-ioral synthesis of ASICs. *IEEE Transactions on Computer-Aided Design of Integrated Circuits and Systems*, 8(6):661–679, June 1989.

[76] Dar-Tzen Peng and Kang G. Shin. Static allocation of periodic tasks with precedence constraints. In *Proceedings of International Conference on Distributed Computing Systems*, pages 190–198, 1989.

[77] Shiv Prakash and Alice C. Parker. SOS: synthesis of application-specific heterogeneous multiprocessor systems. *Journal of Parallel and Dis-tributed Computing*, 16:338–351, 1992.

[78] P. Puschner and Ch. Koza. Calculating the maximum execution time of real-time programs. *The Journal of Real-Time Systems*, 1(2):160–176, September 1989.

[79] Iksoo Pyo and Alvin M. Despain. PDAS: Processor design automation system. In *Proceedings of European Design Automation Conference*, 1993.

[80] Krithi Ramamritham. Allocation and scheduling of complex periodic tasks. In *Proceedings of International Conference on Distributed Computing Systems*, pages 108–115, 1990.

[81] Krithi Ramamritham and John A. Stankovic. Scheduling algorithms and operating systems support for real-time systems. *Proceedings of the IEEE*, 82(1):55–67, January 1994.

[82] Krithi Ramamritham, John A. Stankovic, and Perng-Fei Shiah. Efficient scheduling algorithms for real-time multiprocessor systems. *IEEE Transactions on Parallel and Distributed Systems*, 1(2):184–194, April 1990.

[83] Christopher Rosebrugh and Eng-Kee Kwang. Multiple microcontrollers in an embedded system. *Dr. Dobbs Journal*, pages 48–57, January 1992.

[84] James A. Rowson. Hardware/software co-simulation. In *Proceedings of Design Automation Conference*, pages 439–440, 1994.

[85] K. A. Sakallah, T. N. Mudge, and O. A. Olukotun. Timing verification and optimal clocking of synchronous digital circuits. In *Proceedings of IEEE International Conference on Computer-Aided Design*, 1990.

[86] Lui Sha, Ragunathan Rajkumar, and John P. Lehoczky. Real-time scheduling support in futurebus+. In *Proceedings of Real-Time Systems Symposium*, pages 331–340, 1990.

[87] Lui Sha, Ragunathan Rajkumar, and John P. Lehoczky. Real-time computing with IEEE futurebus+. *IEEE Micro*, June 1991.

[88] Lui Sha, Ragunathan Rajkumar, and Shirish S. Sathaye. Generalized rate-monotonic scheduling theory: A framework for developing real-time systems. *Proceedings of the IEEE*, 82(1):68–82, January 1994.

[89] Kang G. Shin and Parameswaran Ramanathan. Real-time computing: A new discipline of computer science and engineering. *Proceedings of the IEEE*, 82(1):6–24, January 1994.

[90] Mani B. Srivastava and Robert W. Brodersen. SIERA: A unified framework for rapid-prototyping os system-level hardware and software. *IEEE Transactions on Computer-Aided Design of Integrated Circuits and Systems*, CAD-14(6):676–693, June 1995.

[91] Ashok Sudarsanam and Sharad Malik. Memory bank and register allocation in software synthesis for ASIPs. In *Proceedings of IEEE International Conference on Computer-Aided Design*, pages 388–392, 1995.

[92] Andrés Takach, Wayne Wolf, and Miriam Leeser. An automaton model for scheduling constraints. *IEEE Transactions on Computers*, 44(1):1–12, January 1995.

[93] Markus Theibinger, Paul Stravers, and Holger Veit. Castle: An interactive envioronment for hw-sw co-design. In *Proceedings of International Workshop on Hardware-Software Co-Design*, pages 203–210, 1994.

[94] Peter Thoma. Future needs for automotive electronics. In *Proceedings of Design Automation Conference*, pages 72–77, 1994.

[95] Donald E. Thomas, Jay K. Adams, and Herman Schmit. A model and methodology for hardware-software codesign. *IEEE Design & Test of Computers*, 10(3):6–15, September 1993.

[96] Frank Vahid, Jie Gong, and Daniel D. Gajski. A binary-constraint search algorithm for minimizing hardware during hardware/software partitioning. In *Proceedings of European Design Automation Conference*, pages 214–219, 1994.

[97] Frank Vahid, Sanjiv Narayan, and Daniel D. Gajski. SpecCharts: A VHDL front-end for embedded systems. *IEEE Transactions on Computer-Aided Design of Integrated Circuits and Systems*, CAD-14(6):694–706, June 1995.

[98] P. Vanbekbergen, G. Goossens, and H. De Man. Specification and analysis of timing constraints in signal transition graphs. In *Proceedings of the European Conference on Design Automation*, pages 302–306, 1992.

[99] J. Varghese, M. Butts, and J. Batcheller. An efficient logic emulation system. *IEEE Transactions on VLSI Systems*, 1(2):171–174, June 1993.

[100] Elizabeth A. Walkup and Gaetano Borriello. Interface timing verification with application to synthesis. In *Proceedings of Design Automation Conference*, pages 106–112, 1994.

[101] Wayne Wolf. Guest editor's introduction: Hardware-software codesign. *IEEE Design & Test of Computers*, 10(3):5, September 1993.

[102] Wayne Wolf. Architectural co-synthesis of distributed embedded computing systems. Submitted to IEEE Transactions on VLSI Systems, 1994.

[103] Wayne Wolf. Hardware-software co-design of embedded systems. *Proceedings of the IEEE*, 82(7):967–989, July 1994.

[104] Wayne Wolf and Richard Manno. High-level modeling and synthesis of communicating processes using vhdl. *IEICE Transactions on Information and Systems*, E76-D(9):1039–1046, September 1993.

[105] Wayne H. Wolf and Alfred E. Dunlop. Symbolic layout and compaction. In Bryan T. Preas and Michael J. Lorenzetti, editors, *Physical Design Automation of VLSI Systems*, chapter 6, pages 211–281. Benjamin-Cummings, 1988.

[106] W. Ye, Rolf Ernst, Thomas Benner, and Jörg Henkel. Fast timing analysis for hardware-software co-synthesis. In *Proceedings of IEEE International Conference on Computer Design*, pages 452–457, 1993.

[107] Ti-Yen Yen, Alex Ishii, Al Casavant, and Wayne Wolf. Efficient algorithms for interface timing verification. In *Proceedings of European Design Automation Conference*, pages 34–39, 1994.

[108] Ti-Yen Yen, Alex Ishii, Al Casavant, and Wayne Wolf. Efficient algorithms for interface timing verification. submitted to International Journal of Formal Methods, 1995.

[109] Ti-Yen Yen and Wayne Wolf. Optimal scheduling of finite-state machines. In *Proceedings of IEEE International Conference on Computer Design*, pages 366–369, 1993.

[110] Ti-Yen Yen and Wayne Wolf. Communication synthesis for distributed embedded systems. In *Proceedings of IEEE International Conference on Computer-Aided Design*, pages 288–294, 1995.

[111] Ti-Yen Yen and Wayne Wolf. Performance estimation for real-time distributed embedded systems. In *Proceedings of IEEE International Conference on Computer Design*, pages 64–69, 1995.

[112] Ti-Yen Yen and Wayne Wolf. Performance estimation for real-time distributed embedded systems. submitted to IEEE Transactions on Parallel and Distributed Systems, 1995.

[113] Ti-Yen Yen and Wayne Wolf. Sensitivity-driven co-synthesis of distributed embedded systems. In *Proceedings of 8th International Symposium on System Synthesis*, 1995.

[114] Ti-Yen Yen and Wayne Wolf. An efficient graph algorithm for FSM scheduling. *IEEE Transactions on VLSI Systems*, 4(1):98–112, March 1996.

[115] Rumi Zahir. *Controller Synthesis for Application Specific Integrated Circuits*, volume 16 of *Series in Microelectronics*. Hartung-Gorre Verlag, Konstanz, Germany, 1991.

INDEX

Allocation, 5, 50, 54, 58, 87, 118, 126

Application process, 53, 55, 116, 121, 123

Application software architecture, 48

Application specific, 4, 23, 51, 115

ASIC, 3

ASIP, 2

Bounded, 19, 37, 41, 45, 57, 69

Bus scheduling, 21, 55, 126

Bus, 52

Communication link, 18, 33, 49–50, 53, 115

Communication process, 53–54, 116–117, 119, 126–127

Communication time, 53, 117, 121

Computation time, 14, 41–43, 45, 49, 51, 57–58, 81, 99, 117

Control-dominated, 23, 26, 48

Core-based design, 3

Data dependency, 17, 19, 43, 54, 57, 62

Deadline, 6, 14, 45, 48, 58, 80, 87–88, 90

DelayEstimate, 76, 80

Design goal attribute, 87

Distributed system, 9, 18, 86, 135

Dummy process, 45–46, 50, 64, 80, 84

EarliestTimes, 71, 75–76

Embedded system architecture, 48, 58, 95, 98, 103

Emulation, 3

END node, 44, 63–64

Engine metaphor, 4

Fixed-point iteration, 60, 64, 67, 123, 135

Fractional deadline, 90, 93, 126

Hard cost constraint, 49, 87, 97, 104, 129

Hard deadline, 44–45, 87, 90, 98, 126

Hardware engine, 4, 9, 48–49, 58

Idle-PE elimination, 94, 96–97, 99, 103

Inter-task communication, 47, 129

Inverse-deadline priority assignment, 17, 90, 126

Latest finish time estimate, 64, 66

Latest finish time, 64, 90

Latest request time estimate, 64, 66

Latest request time, 64, 90

Latest required time, 90, 126

LatestTimes, 64, 66–67, 69–71, 75–76

LCM approach, 18, 59, 81, 135

Levels of abstraction, 1–2

Load-balancing, 94, 99, 103

Max constraint, 37, 69

Maximum separation, 37, 69, 71

MaxSeparations, 71, 74–76

Message, 21–22, 53–54, 115, 117, 126–127, 138

PE downgrading, 98–99, 103

PE type, 42, 49, 54, 98–101

Period, 14, 22, 44–45, 51, 77, 80–81, 90, 99, 137

Periodic, 14, 18, 50–51, 60, 138

156

Phase adjustment, 62, 64, 75, 77,
    135
Point-to-point communication, 19,
    51, 115, 118, 136
Preemption, 51, 55, 57, 64, 68, 86,
    123, 135
Priority scheduling, 15, 51, 54, 60,
    136
Process cohesion, 50, 99–100
Process exclusion, 50, 100
Process, 41
Processing element, 48
Processor utilization, 9, 15, 60, 62,
    94
Rate constraint, 27, 44, 86
Rate-monotonic scheduling, 8, 14,
    22, 90
Reactive system, 6
Real time, 5
Receiving process, 46, 53, 117, 121
Release time, 17, 45
Response time, 13, 17, 42, 44, 60,
    119, 123
Sending process, 46, 53, 117, 121
Sensitivity, 87, 100, 127, 136
Single bus, 19, 115
Soft cost constraint, 49, 87, 101,
    129
Soft deadline, 44–45, 87, 101, 129
START node, 44, 63
System cost, 49, 54, 87, 93–94, 115
System on silicon, 3
System-level ASIC, 3
Task graph, 41, 43
Task, 43
Uninterrupted execution, 13, 41,
    57
Worst-case task delay, 44